# AGRICULTURAL EXPANSION and Tropical Deforestation

# AGRICULTURAL EXPANSION
# and Tropical Deforestation

## Poverty, International Trade and Land Use



**Solon L Barraclough and Krishna B Ghimire**

Earthscan Publications Ltd, London and Sterling, VA

First published in the UK and USA in 2000 by
Earthscan Publications Ltd



A catalogue record for this book is available from the British Library

ISBN: 1 85383 665 6 paperback
1 85383 666 4 hardback

Typesetting by Composition & Design Service
Printed and bound by Creative Print and Design (Wales)
Cover design by Susanne Harris

For a full list of publications please contact:

Earthscan Publications Ltd
120 Pentonville Road
London, N1 9JN, UK
Tel: +44 (0)20 7278 0433
Fax: +44 (0)20 7278 1142
Email: earthinfo@earthscan.co.uk
http://www.earthscan.co.uk

22883 Quicksilver Drive, Sterling, VA 20166–2012, USA

Earthscan is an editorially independent subsidiary of Kogan Page Ltd and publishes in association with WWF-UK and the International Institute for Environment and Development

This book is printed on elemental chlorine-free paper

# CONTENTS

# List of Maps, Tables and Boxes

## Maps



## Tables



## Boxes

# About the Cooperating Organizations

## UNRISD

The United Nations Research Institute for Social Development (UNRISD) is an autonomous agency engaging in multi-disciplinary research on the social dimensions of contemporary problems affecting development. Its work is guided by the conviction that, for effective development policies to be formulated, an understanding of the social and political context is crucial. UNRISD attempts to provide governments, development agencies, grassroots organizations and scholars with a better understanding of how development policies and processes of economic, social and environmental change affect different social groups. Working through an extensive network of national research centres, UNRISD aims to promote original research and strengthen research capacity in developing countries.

Current research programmes include: Civil Society and Social Movements; Democracy and Human Rights; Identities, Conflict and Cohesion; Social Policy and Development; and Technology and Society.

A list of free and priced publications available from UNRISD can be obtained by contacting the Reference Centre, UNRISD, Palais des Nations, 1211 Geneva 10, Switzerland; Tel (41 22) 917 3020; Fax (41 22) 917 0650; Telex 41.29.62 UNO CH; Email: info@unrisd.org; Website: http://www.unrisd.org.

## WWF-International

In just over three decades, the World Wide Fund For Nature (WWF), formerly known as the World Wildlife Fund, has become the world's largest and most respected independent conservation organization. With almost 5 million supporters distributed throughout five continents, 24 national organizations, five associates and 26 programmes, WWF can safely claim to have played a major role in the evolution of the international conservation movement.

Since 1985, WWF has invested over US$1,165 million in more than 11,000 projects in 130 countries. All these play a part in the campaign to stop the accelerating degradation of Earth's environment, and to help its human inhabitants live in greater harmony with nature.

For more information WWF-International can be contacted at Avenue du Mont Blanc, CH-1196, Gland, Switzerland; Tel (41 22) 364 9111; Fax (41 22) 364 5358; Website: http://www.panda.org.

# List of Acronyms and Abbreviations

| | |
|---|---|
| CDC | Cameroon Development Corporation |
| CEPAL | Comision Económica para América Latina y el Caribe |
| CONAP | National Council for Protected Areas (Brazil) |
| DIGEPOS | National Forest Service (Brazil) |
| EU | European Union |
| FAO | Food and Agriculture Organization of the United Nations |
| FDN | La Fundación Defensores de la Naturaleza (Guatemala) |
| FELDA | Federal Land Development Authority (Malaysia) |
| FUNAI | National Indian Foundation (Brazil) |
| FYDEP | Empresa de Fomento y Desarrollo de Petén (Agency for Development Promotion in Petén) (Guatemala) |
| GDP | gross domestic product |
| GVIAO | gross value of industrial and agricultural output |
| INCRA | National Institute for Colonization and Agrarian Reform (Brazil) |
| INPE | Instituto Nacional de Pesquisas Espaciais (National Institute for Space Research, Brazil) |
| INTA | National Agrarian Transformation Institute (Guatemala) |
| ITTO | International Tropical Timber Organization |
| KPD | Rural Development Corporation (Malaysia) |
| MAB | Man and Biosphere Programme (UNESCO) |
| MAIs | multinational agreements on investments |
| MIA | Missão Anchieta (Brazil) |
| MIDENO | North-West Development Authority (Cameroon) |
| MOA | Ministry of Agriculture (China) |
| NGOs | non-governmental organizations |
| OECD | Organisation for Economic Co-operation and Development |

| | |
|---|---|
| QGLYTJHB | Quanguo Linye Tongji Huibian (National Compendium of Forestry Statistics) (China) |
| UNDP | United Nations Development Programme |
| UNESCO | United Nations Educational, Scientific, and Cultural Organization |
| USAID | United States Agency for International Development |
| WRI | World Resources Institute |
| WTO | World Trade Organization |
| WWF | World Wide Fund For Nature International |
| ZGTJNJ | Zhongguo Tongji Nianjian (China Statistical Yearbook) |

# Study Countries: Maps

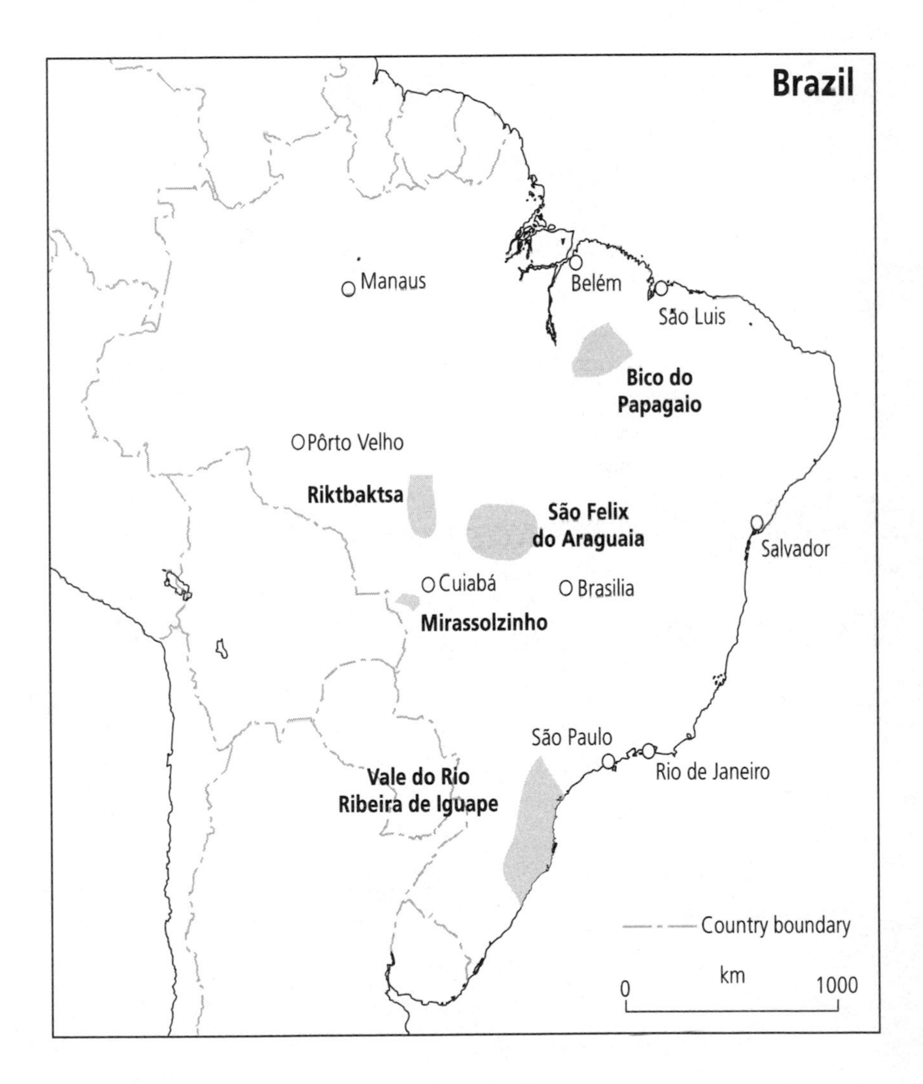

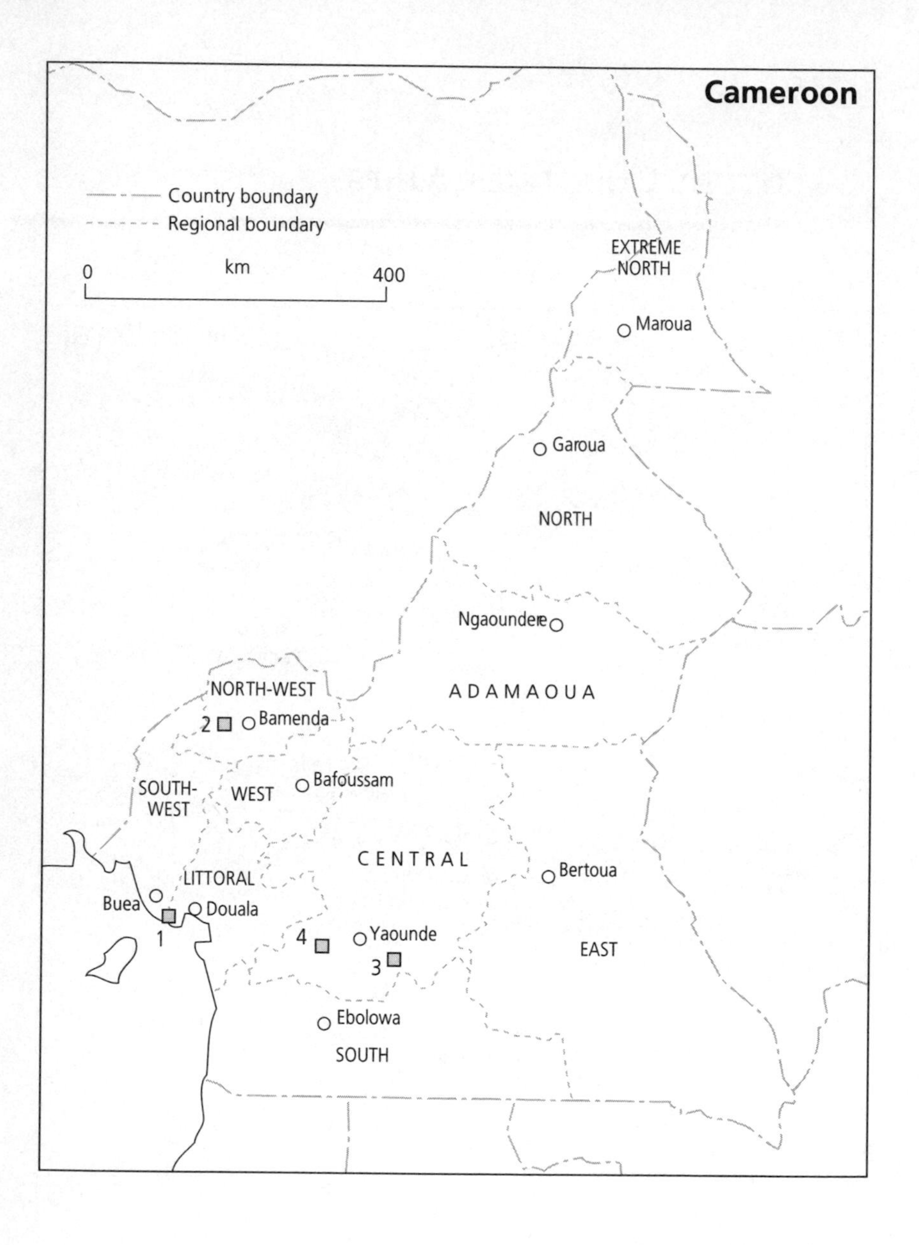
Cameroon
Country boundary
Regional boundary
0
km
400
EXTREME NORTH
Maroua
Garoua
NORTH
Ngaoundere
ADAMAOUA
NORTH-WEST
2
Bamenda
SOUTH-WEST
WEST
Bafoussam
CENTRAL
LITTORAL
Buea
Douala
1
Bertoua
4
Yaounde
3
EAST
Ebolowa
SOUTH

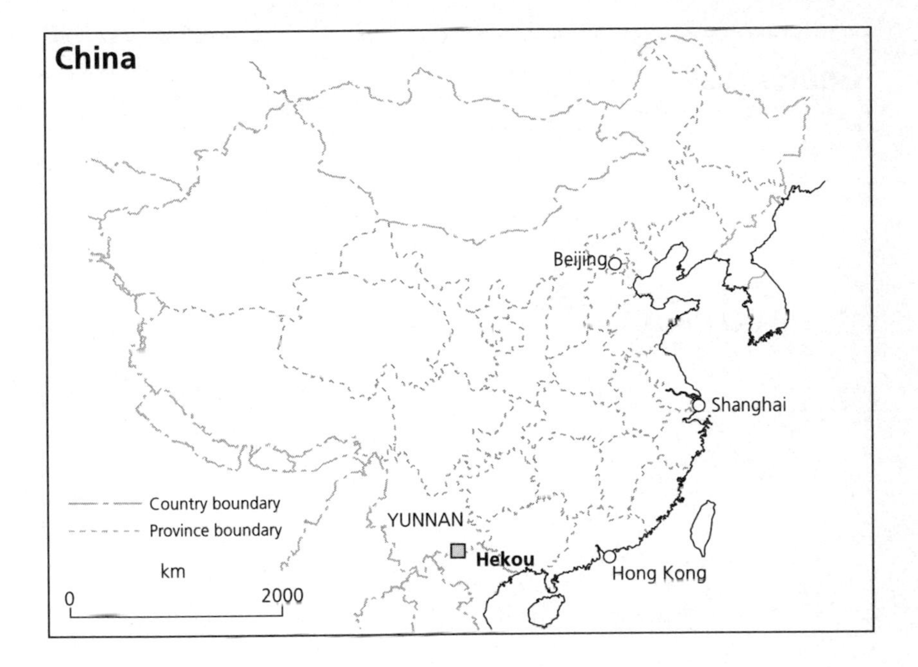
China
Beijing
Shanghai
Country boundary
Province boundary
YUNNAN
Hekou
Hong Kong
km
0
2000

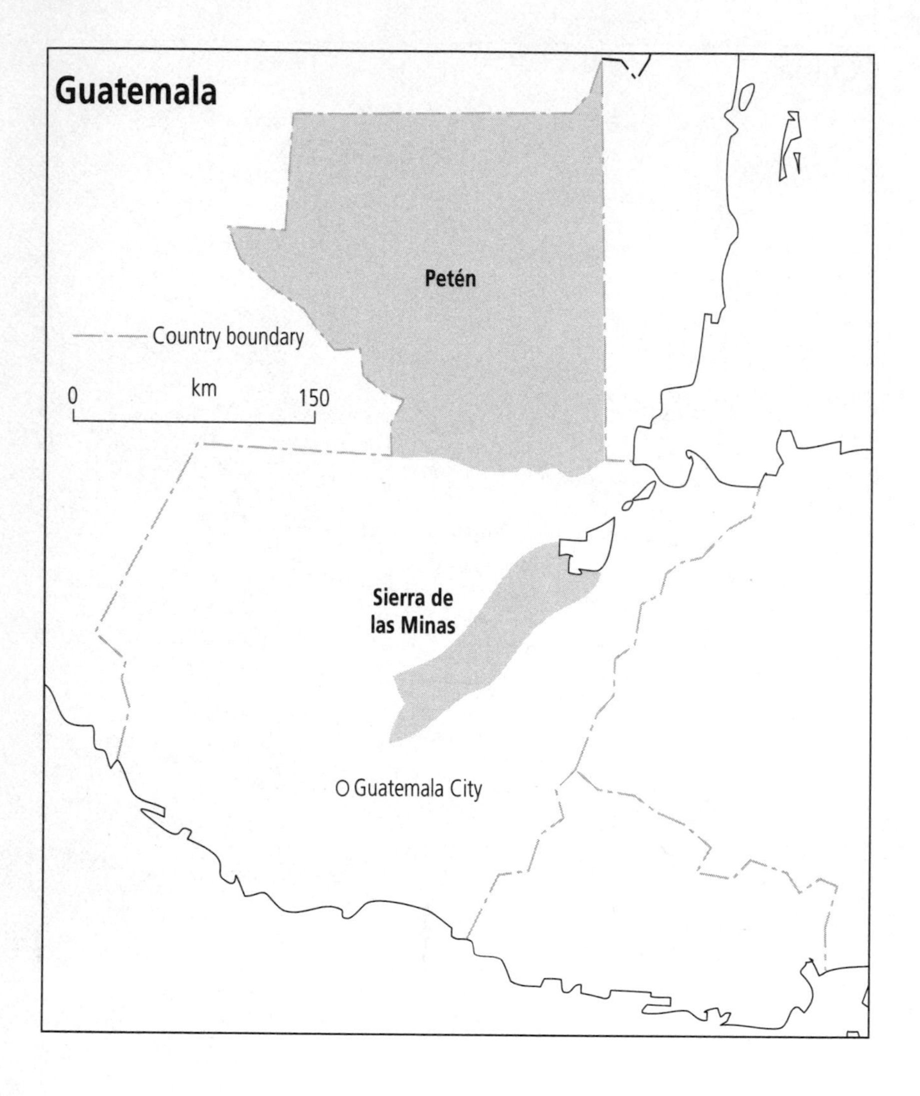
Guatemala
Petén
Country boundary
0
km
150
Sierra de
las Minas
Guatemala City

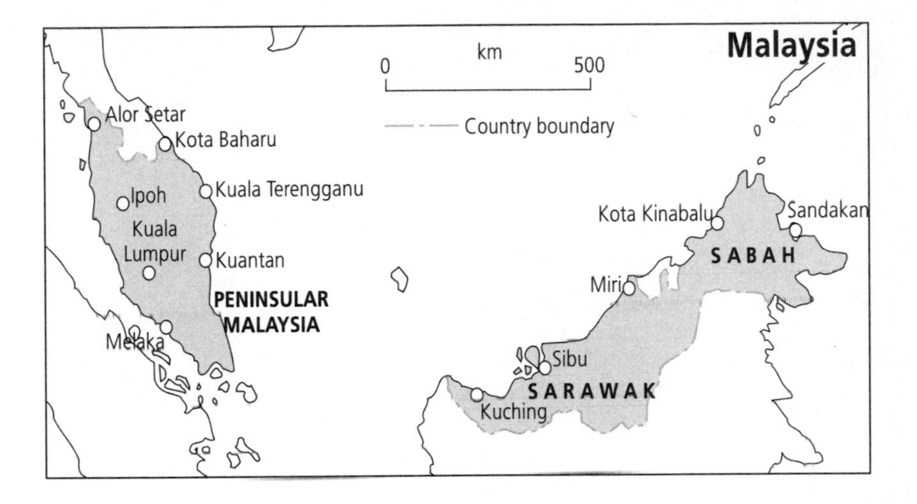
Malaysia
km
0
500
Country boundary
Alor Setar
Kota Baharu
Kuala Terengganu
Ipoh
Kuala Lumpur
Kuantan
PENINSULAR MALAYSIA
Melaka
Kota Kinabalu
Sandakan
SABAH
Miri
Sibu
SARAWAK
Kuching

# 1 INTRODUCTION: SOCIAL DETERMINANTS OF DEFORESTATION

Accelerated tropical deforestation during recent decades has resulted in the conversion of hundreds of millions of hectares of tropical forests to other land uses such as growing crops, pastures, roads, mines, reservoirs, industrial, residential or administrative areas and wastelands. In 1980, about one-tenth of the world's nearly 2 billion hectares of remaining tropical forests were estimated to have been converted to other land uses during the subsequent decade alone. Even vaster areas of tropical forests have been badly degraded by logging, excessive fuelwood extraction, industrial pollution, overgrazing, destructive man-made fires and many other deforestation processes caused by humans.

This version of the deforestation narrative is, however, essentially tautological. It merely views volatile combinations of several long-recognized deforestation processes as being the proximate causes of tropical deforestation. This suggests little in the way of cures other than to attempt to stop such processes.

At a more general level, by definition, human-induced (anthropogenic) deforestation is ultimately caused by people and their activities. The implicit remedies embodied in the definition are to halt or reverse population growth and to eliminate activities stimulating tropical deforestation. Population stabilization or reduction is at best a long-term proposition. Most people entering the labour force and reaching reproductive age during the coming two decades are already with us. Excluding a massive demographic catastrophe, the world population will increase by between one-fourth to one-half by the year 2025, with the majority of this population growth taking place in poor countries. Modifying human activities to become more environmentally friendly appears more promising in terms of stopping destructive deforestation during the foreseeable future, unless one believes that it is beyond human control. Solutions have to be sought to change humankind's values, social relations and activities. This implies the reform of institutions and policies at sub-national, national and international levels. Research can help to indicate how they might be reformed to encourage the socially and ecologically sustainable use of natural resources in tropical forest regions.

Deforestation 'stories' that indicate feasible remedies have to include the social origins of deforestation processes and their social impacts. These social determinants are primarily institutions (relatively stable rules and customs regulating social relations), policies (purposeful courses of action by diverse social actors), and technologies (the applications of science and experience for socially defined practical ends). Deforestation narratives have to confront issues of power relations at all levels from the local to the global. They need to deal with the perceived identities and goals of diverse social actors as well as with the frequently unanticipated and unintended outcomes of conflicting policies pursued in a context of interacting dynamic social and ecological systems and sub-systems. In other words, deforestation stories that could contribute to the more sustainable use of natural resources in tropical forest regions have to include the political and socio-economic dimensions as well as the ecological ones.

This is widely recognized at the conceptual level, but the complexities and uncertainties inherent in analyses of interacting social and natural systems are frequently forgotten in practice. Many researchers are under tremendous pressures to be 'policy relevant' and to recommend 'practical', albeit simplistic, solutions. For example, agricultural expansion together with the plundering of remaining forests is spurred by population growth and trade. These are widely viewed as the leading proximate and 'root' sources of recent tropical deforestation. This may sometimes be the case. If true, what does it imply in the way of policy and institutional changes locally, nationally and internationally? Each local and national situation is, to some extent, unique and constantly changing. What reforms could be effective and feasible in different places and times? The international context is also constantly and often dramatically changing. What international reforms could, in the current context, contribute to more sustainable tropical forest resource management?

These questions are not answerable in any definitive manner, but they guided our research. This volume summarizes some of the findings. It argues that deforestation is an outcome of policies pursued by diverse social actors within interacting social and ecological systems at local, national and international levels. In other words, socially undesirable tropical deforestation is a systemic problem that requires deep policy and institutional reforms at all levels. The research shows that agricultural expansion and international trade are important factors but that their roles are varied and frequently contradictory. Their impact on livelihoods and on tropical deforestation depends largely on the contexts in which they occur. In this book we try to sort out several of the complex linkages between the policies of governments and of other social actors on the one hand, and social structures on the other. Finally, we look at public policy and institutional reforms at different levels that could help to promote more sustainable uses of tropical forest resources.

## A COMPLEX ISSUE

Tropical deforestation has been a major theme in the countless discussions, reports and publications leading to the 1992 United Nations Conference on Environment and Development in Rio. It will undoubtedly continue to be a central international environmental issue during coming years. However, there is much disagreement, even among specialists, about the dynamics of deforestation and its socioeconomic and ecological implications if the widely conflicting claims about its causes, extent, impacts and remedies can be taken as evidence. This is particularly true of the role of agricultural expansion that is frequently blamed for some 60 per cent of the current rapid deforestation in the tropics of up to 20 million hectares annually (World Bank, 1992).

The reasons why shrinking areas of tropical forests arouse increasing anxiety are now widely known. The livelihoods of over 200 million forest dwellers and poor settlers depend directly on food, fibre, fodder, fuel and other resources taken from the forest or produced on recently cleared forest soils. Many millions more live from employment in forest based crafts, industries and related activities. Numerous indigenous groups are threatened with genocide induced by alienation or destruction of their source of life support. Degradation of forest habitats is accompanied by the extinction of many species of flora and fauna. This loss of biodiversity poses fundamental ethical questions as well as more material ones about lost options for the future. Ecosystems upon which humans ultimately depend may collapse. Soil erosion, salinization and compaction may prove irreversible, as may adverse changes in local and regional climates. Deforestation is frequently accompanied by more devastating floods downstream and the depletion of water reserves in underground aquifers, lakes and reservoirs. Tropical deforestation contributes to the build-up of greenhouse gases that may induce global climate change with incalculable consequences. Future supplies of food, fuel and timber to meet the needs generated by economic growth and increasing populations could be imperilled or become more costly. Rapidly expanding mass tourism in many poor countries poses both new threats and opportunities for more sustainable uses of tropical forests. Conflicts of interests between transnational corporations mostly based in the North and those of many rural poor in developing countries are intensifying, as are conflicts between rich and poor country governments about the proper management of 'the heritage of all mankind'.

Not all deforestation is incompatible with sustainable development. The world's temperate forests have been reduced by over one-third in recent centuries. Many of these formerly forested areas were cleared for agricultural and other human uses. Much of this former forest land now supports large and relatively prosperous populations with highly productive farms, industries and cities. Deforestation has apparently stabilized in industrialized countries. There are also large once-forested areas in the tropics that have supported dense populations for centuries.

People denied other alternatives than wresting a bare living by clearing forest will try to survive even where conditions render continuous cultivation unsustainable. At the same time, lucrative short-term profits can frequently be reaped by powerful élites in both industrialized and developing countries. Northern investors and consumers commonly benefit disproportionately from cash crop and timber exports from the tropics at the expense of forest-dependent poor people in the South and a sustainable environment.

Many interrelated processes contribute to tropical deforestation. Agricultural expansion is prominent among them, but this in turn includes numerous sub-processes responding to different dynamics. Moreover, local level deforestation processes differ greatly from place to place and over time. Simplistic generalizations based on global or regional and national data are not very helpful in understanding the complex causes and social and ecological impacts of deforestation, or in suggesting remedial actions, in specific localities. An analytical case study approach is more appropriate. This was shown by the authors' earlier research into the social dynamics of deforestation (Barraclough and Ghimire, 1995).

## PRINCIPAL QUESTIONS GUIDING THE RESEARCH

The research first reviewed estimates by the Food and Agriculture Organization (FAO) of the United Nations and a number of other international sources of recent land use changes and deforestation trends in developing countries. Case studies were then commissioned in five countries – Brazil, Guatemala, Cameroon, Malaysia and China. The objective of the case studies was to explore critically the dynamics of tropical deforestation in specific socioeconomic, political and ecological contexts. Special emphasis was placed on the roles of agricultural expansion and international trade in stimulating deforestation processes.

The research followed a political economy approach. It attempted to identify the nature and importance of diverse socioeconomic processes leading to tropical deforestation in specific subnational regions and localities. It gave explicit attention to the social actors involved and on how they may have benefited or been negatively affected by the clearance or degradation of tropical forests and by the expansion of agriculture into forest areas. The implications of these processes for the livelihoods of low-income groups directly or indirectly affected was another principal focus of the case studies. This identification and analysis of deforestation processes constituted the first set of research issues and questions.

The second set of issues related to the roles of policies in stimulating, directing or checking deforestation as well as in magnifying or attenuating its social and ecological impact. Public policies at national levels are always crucial, but sub-national and international policies of the state and of other social actors, such as corporate bodies and NGOs, can also be extremely important.

Governments are only one component of the institutional framework that defines and regulates any society. Policy analysis is rather meaningless, especially for comparative purposes, unless it is carefully linked with a society's broader institutional framework in which government policies are generated and carried out. The institutional determinants of deforestation processes constitute, a third cluster of issues and questions. The research considered both policies and institutions, emphasizing their linkages and dynamic interactions.

In the policy field, land use and agricultural and forest policies should obviously receive attention. Policies regarding land settlement, rural development and forest protection readily come to mind. Many policies that may appear remote from deforestation processes can often be crucial. Price and trade policies, fiscal policies and those affecting employment and welfare frequently contribute to accelerating tropical deforestation, as do consumption and production patterns in both rich industrial countries and poor agrarian ones.

The political institutions of government at all levels, as well as economic institutions organizing and regulating production, trade and consumption, mediate policies and market forces. Logging, mining, infrastructure construction and agricultural expansion often directly drive deforestation processes. Both land tenure and farming systems are closely related institutions that were central everywhere in this research. In certain circumstances, however, population movements and environmental dynamics have also played an important role.

### *Land tenure*

Land tenure institutions determine the rights and obligations of different social actors, such as individuals, clans, local communities, corporate bodies and the state in access to land, water, forests and other natural resources and in the distribution of their benefits. In agrarian societies land tenure also defines the obligations of those who work the land in relation to those who accumulate its surplus. Analyses of land tenure systems are central for understanding the nature of tropical deforestation processes, who benefits from them and who is prejudiced. Land tenure relationships are a good indicator of social relationships in the broader society. They reflect the relative power of different social classes and diverse ethnic groups. Land tenure systems in distant non-forested areas may frequently be a principal factor in forcing landless workers and peasants to invade tropical forests.

### *Farming systems*

Farming systems constitute distinctive combinations of social relations (for example, land tenure), farming practices and technologies, land use and cropping patterns, consumption standards, access to markets and so on, that tend to go together and reproduce themselves. They reflect both social relations and economic structures of the broader society as well as the constraints imposed on agriculture by climate, soils, water availability and biological endowments. Very different farming systems can coexist in the same ecological context. Low external input systems aimed primarily at self-provisioning and high external input systems producing for national or international markets are often found side by side. Within each broad type of farming system there can be numerous sub-systems with distinctive social relations, land use and production patterns.

### *Demographic issues*

Demographic issues raise a further cluster of complex research questions. In-migration, out-migration, birth rates and mortality rates all interact among themselves as well as with political, socioeconomic and environmental factors. Their impact on agricultural expansion and tropical deforestation has to be analysed in each unique context, as broad generalizations can be very misleading.

### *Natural environment*

The same is true of the constraints imposed by the natural environment. Soils, climate, water availability and the biological dynamics of each ecosystem have to be taken into account. One does not cultivate water-demanding crops in areas of semi-arid savannah tropical forests without encountering great difficulties, while dryland crop production in areas of humid tropical forests is seldom an attractive proposition. These constraints influence both agricultural expansion and possible alternatives.

### *Alternatives*

Finally, what are the alternatives to tropical forest clearance for people depending upon agriculture for their livelihoods? Much depends upon the level of analysis. Alternatives for residents of a local ecosystem such as a particular river basin, plateau or community may appear very limited, but they become much less constrained if broader ecological and political boundaries are assumed. We have attempted to look at some of the initiatives to check agricultural expansion into tropical forests at various levels from local communities to the nation state and beyond.

## THE CASE STUDY COUNTRIES

### *Brazil*

Brazil includes a major portion of the Amazon rain forests that are threatened by agricultural expansion and other deforestation processes as well as several other tropical forest areas that have already been largely cleared. According to the FAO, 36,780 million hectares of new land were brought into agricultural production between 1973 and 1992 (Table 1.1), but with a dramatic reduction in forest and woodland areas, amounting to over 100 million hectares. This was nearly three times the area apparently lost to agricultural expansion. Much of the land cleared of forests and woodlands became wasteland or went into other uses. 'Other land' had increased by some 65,000 hectares. During this period Brazil experienced a rapid increase in road building, dam construction, mining and urbanization. Indeed, the urban proportion of the country's population increased from 50 per cent of the total population to 75 per cent between 1965 and 1990, while its rural population diminished in absolute numbers (UNDP, 1994).

Recent national level estimates of deforestation, based on satellite images, indicate that during the 1980s deforestation in the

**Table 1.1** *Agricultural Expansion and Deforestation in Case Study Countries*

| | *1973* | *1979* | *1989* | *1992* | *Difference between 1973–1992 (in 1000ha)* | *(in %)* |
|---|---|---|---|---|---|---|
| **Brazil** | | | | | | |
| Total area | 851,197 | 851,197 | 851,197 | 851,197 | – | – |
| Land area | 845,651 | 845,651 | 845,651 | 845,651 | – | – |
| Arable–permanent crops | 57,820 | 68,970 | 78,650 | 59,000 | 1180 | 2 |
| Permanent pasture | 151,200 | 160,000 | 170,000 | 186,800 | 35,600 | 24 |
| Forest–woodland | 589,850 | 577,430 | 553,130 | 488,000 | -101,850 | -17 |
| Other land | 46,781 | 39,251 | 43,871 | 111,851 | 65,070 | 139 |
| **Guatemala** | | | | | | |
| Total area | 10,889 | 10,889 | 10,889 | 10,889 | – | – |
| Land area | 10,843 | 10,843 | 10,843 | 10,843 | – | – |
| Arable–permanent crops | 1613 | 1726 | 1875 | 1885 | 272 | 17 |
| Permanent pasture | 1230 | 1290 | 1390 | 1420 | 190 | 15 |
| Forest–woodland | 5010 | 4630 | 3830 | 3590 | -1420 | -28 |
| Other land | 2990 | 3197 | 3748 | 3948 | 958 | 32 |
| **Cameroon** | | | | | | |
| Total area | 47,544 | 47,544 | 47,544 | 47,544 | – | – |
| Land area | 46,540 | 46,540 | 46,540 | 46,540 | – | – |
| Arable–permanent crops | 6160 | 6912 | 7008 | 7040 | 880 | 14 |
| Permanent pasture | 8300 | 8300 | 8300 | 8300 | – | – |
| Forest–woodland | 26,400 | 25,750 | 24,650 | 24,330 | -2070 | -8 |
| Other land | 5680 | 5578 | 6582 | 6870 | 1190 | 21 |
| **Malaysia** | | | | | | |
| Total area | 32,975 | 32,975 | 32,975 | 32,975 | – | – |
| Land area | 32,855 | 32,855 | 32,855 | 32,855 | – | – |
| Arable–permanent crops | 4580 | 4765 | 4880 | 4880 | 300 | 7 |
| Permanent pasture | 26 | 27 | 27 | 27 | 1 | 4 |
| Forest–woodland | 22,940 | 21,500 | 19,100 | 19,352 | -3588 | -16 |
| Other land | 5309 | 6563 | 8848 | 8596 | 3287 | 62 |
| **China** | | | | | | |
| Total area | 959,696 | 959,696 | 959,696 | 959,696 | – | – |
| Land area | 932,641 | 932,641 | 932,641 | 932,641 | – | – |
| Arable–permanent crops | 101,376 | 100,415 | 96,115 | 96,302 | -5074 | -5 |
| Permanent pasture | 319,080 | 319,080 | 319,080 | 400,000 | 80,920 | 25 |
| Forest–woodland | 113,624 | 136,365 | 126,465 | 130,495 | 16,871 | 15 |
| Other land | 398,561 | 376,781 | 390,981 | 305,844 | -92,717 | -23 |

*Source: FAO Yearbooks, 1973–1993*

country was taking place at a rate of nearly 3 million hectares annually. Much of the forest clearance was concentrated in southern and south-eastern peripheral sub-regions such as Mato Grosso, Goiás, Rondônia and Pará. Rapid agricultural modernization in the south in the 1970s had left many farm workers and peasants without jobs or land. The government's policies were to settle as many as possible of these and other land-seeking people in the 'empty' forested areas of the Amazon. The state also provided tax credits and other fiscal incentives for large-scale agriculture, cattle ranching, logging and mining. These policies can be viewed as principal causes of extensive deforestation.

In order to analyse relations between patterns of agricultural expansion, public policies and the destruction of forests, as well as associated impacts on the livelihoods of local populations, five local-level case studies were carried out. Four of these studies were undertaken inside or within the periphery of the Amazon region, and the remaining one was conducted in a south-eastern forest area that included remnants of its once extensive Atlantic coastal forests (Mata Atlantica).

One study area was the Bico de Papagaio region in the state of Maranhao. This area has been subject to immigration by successive waves of poor settlers, and also to intense land speculation, since the opening of the Belém–Brasilia highway in the 1960s. A second study was carried out in the Mirassolzinho area of western Mato Grosso where squatters, cattle ranchers and government-supported agricultural corporations have been active in forest clearance. A third focused on Kilakta Indians' use of natural resources and their conflicts with outsiders in two Indian reserve areas in north-western Mato Grosso. A fourth study area was the São Felix do Araguaia area of north-eastern Mato Grosso, where government-financed corporations and large landholders have been active in clearing extensive forest areas for pasture and beef production. The fifth study was on the valley of Rio Ribeira de Iguape (in south-eastern São Paulo state bordering the state of Paraná) where nearly half of the remaining Atlantic coastal rain forests are found; these forests are threatened by various deforestation processes and are also the focus of numerous NGO and government conservation initiatives.

### *Guatemala*

Deforestation in Guatemala over the last 20 years has been particularly rapid. The FAO estimates a decline in forest and woodland coverage between 1973 and 1992 from 5010 million hectares to 3590 million hectares, or by nearly one-third (Table 1.1). Agricultural area has also expanded, but much less than the decline in forest area. The remaining loss of the forest area was accounted for by the increase in 'other land', in part for urbanization and in part for non-agricultural uses including barren lands.

The study in Guatemala focused on the impact of local-level production systems on deforestation and the linkages between agricultural expansion and forest clearance in two of the country's major forest regions. It also sought to examine, based mainly on the review of the available literature, the impact of international commodity markets, trade and foreign aid on agricultural and deforestation processes. It attempted to show how government agricultural policies since the mid-1950s have been based on the economic logic of profit maximization, with little attention given to the protection of forests or the livelihood of the peasants.

One case study was carried out in the north-eastern Petén. This region includes nearly half the country's remaining forests. It has been a principal focus of government road construction and colonization programmes since the 1960s. It was also the site of several recent environmental initiatives. The second region was the little-studied north-eastern La Sierra de la Minas region. It includes a wide variety of ecological conditions and of Indian and *ladino* (people of mixed European and indigenous descent) smallholder settlements as well as large estates known as *latifundia*. The remaining forested areas in this region are in imminent danger of destruction and it has been a site of numerous conflicts.

### *Malaysia*

Forests and woodlands dwindled most rapidly in Malaysia during the 1950s and 1960s. Between 1973 and 1992, forested area declined more slowly from 22,940 million hectares to 19,352 million hectares (Table 1.1). About one-third of this area went to agriculture, but the remaining two-thirds went to other uses. This suggests that the forest areas that were cleared were either turned into barren lands or were used for urbanization and infrastructure. The urban population nearly doubled between 1960 and 1993 (UNDP, 1994). Deforestation associated with agricultural expansion has virtually ceased in Peninsular Malaysia. In Sarawak and Sabah commercial logging for export, especially to Japan, has been a primary process directly generating deforestation in recent years.

The Malaysian case study covered the country's three principal regions: Peninsular Malaysia, Sabah and Sarawak. The study approached the themes of agricultural expansion and forest clearance at three levels by providing a macro overview, three regional surveys and several local-level case studies. The macro overview looked at trends in forest clearance, examined ecological constraints, identified linkages with the world economy and explored available alternatives. The regional surveys considered the situation in Peninsular Malaysia, Sabah and Sarawak separately, because each region differs significantly in terms of history, land use, socioeconomic pressures and institutional organization. The micro case studies examined the agriculture–forest interface at community level.

The government's initial land development programme after independence was aimed, in part, at reducing rural poverty and, in part, at increasing export earnings. It included the development of large plantations to produce rubber and oil palm. These were responsible for much of the deforestation that took place in Peninsular Malaysia. Logging and mining activities have been much more important in the deforestation of Sabah and Sarawak. The development of tourism and aquaculture has also adversely affected forests, especially mangroves, in recent years, in some areas. The country has industrialized rapidly. Primary commodities such as petroleum, timber, oil palm and rubber still remain important export earners although industrial exports have recently become dominant. For Sarawak and Sabah, however, exports of timber are crucial. This has ramifications for deforestation in these states that include most of the country's remaining forests.

### *Cameroon*

Cameroon in the 1980s is believed by some independent scholars to have had the highest rates of deforestation in Central Africa. The

FAO's estimate of forest clearance between 1973–1992 indicates a decline of total forest area of 2,070,000 hectares (Table 1.1). Much of this land has apparently been used for agricultural production, although a significant area has also been incorporated into 'other lands'. As the country experienced a rapid urbanization process during this period, the increase in the 'other land' category appears credible.

The case study in Cameroon examined a number of specific crucial processes related to agricultural expansion and deforestation. These included export-oriented cash crop production driven by the establishment of large para-statal plantations, a rapid rise in export crop production among middle and rich farmers, and commercial logging. Rural inequalities and poverty increased as a result of urban-biased and market-driven public policies that failed to recognize customary land rights.

To elucidate these processes, four case studies were undertaken. The first looked at the impacts of the Cameroon Development Corporation (CDC) agro-industrial plantations on deforestation and livelihoods in South-western province. The second focused on changes and trends in the farming systems in the montane forest ecosystems in the Kilum massif area in Northwest province. The third investigated the processes of forest clearance ensuing from commercial logging and family farming in the Mbalmayo forest reserve area in Central province. The fourth case study examined the social dynamics of deforestation mainly within the traditional peasant farming of the southern Bankundu area in Central province.

### *China*

China still has some rich tropical forest resources and in the past decades there has been an impressive official drive for reafforestation. But there are also indications that logging in the few remaining primary forests has been rapidly advancing (World Bank, 1992). Deforestation problems in ecologically fragile mountains, hills and high plateaux have been documented as contributing to soil erosion, loss of diversity and water shortages (He, 1991; Smil, 1984). Reafforestation efforts have not been as successful as planned (Ross, 1988). Land tenure problems associated with China's recent economic reforms have induced widespread incidents of illegal tree cutting (Menzies and Peluso, 1991). In many areas, there has been continued expansion of agricultural land into forest reserves (Li et al, 1987; Zuo, 1993).

Poverty and population pressures have commonly been linked with the environmental degradation of China's forests (MOA, 1991). Some have referred to the past central planning system and socialist ideology as damaging to the forests. In recent years, the impact of economic liberalization measures is becoming visible for both the expansion of rural industrialization and agricultural expansion.

The study on China focused primarily on local-level agricultural and forest use practices and needs. In particular, a case study was carried out at Hekou county in tropical Yunnan Province to illustrate how problems are manifested in terms of deforestation trends at the local level. This was accompanied by an examination of wider regional and national past and present agricultural trends, and changes in the composition and area of forested land. Recent development strategies and their impact on the forestry sector are especially highlighted.

# 2 The Extent of Tropical Deforestation and Agricultural Expansion in Developing Countries

Land use changes at global and national levels in developing countries in the past, as well as during recent decades, should be examined more closely before later looking at evidence from case studies at sub-national levels. In the mid-1980s about one-third of the Earth's land area was occupied by forests according to the FAO (see Box 2.1 for FAO's forest classifications). A little over half of these forests were considered to be tropical, although they accounted for a slightly lower proportion (about 43

**Box 2.1** *Concepts and Definitions of Different Types of Forests Used in the Tropical Forest Resources Assessment by the FAO*

*Forest:* this is an aggregate to indicate what is normally understood as forest, namely natural forest and forest plantation.
*Closed forest:* stands of broad-leaved (hardwood) forests, which when not recently cleared by shifting agriculture or heavily exploited, cover with their various storeys and undergrowth a high proportion of the ground and do not have a continuous grass layer allowing grazing and the spreading of fires. They are often, but not always, multistoreyed. They may be evergreen, semi-deciduous, or deciduous, wet, moist or dry. As an indication, for remote sensing purposes the crown coverage is 40 per cent or more.
*Open forest:* this refers to mixed broad-leaved forest/grassland formations with a continuous grass layer in which the tree synusia covers more than 10 per cent of the ground.
The division between closed and open hardwood forests is more of an ecological than physiognomic type and is not characterized necessarily by the crown cover percentage. In some woodlands the trees may cover the ground completely, as in closed forests.
The distinction between closed and open forests has not been made for conifers, since it does not have the same ecological importance and is difficult, if not impossible, to apply.

*Shrubs:* any vegetation type where the main woody elements are shrubs (broad-leaved or coniferous species) of more than 50cm and less than 7m in height. The height limits between trees and shrubs should be interpreted with flexibility, particularly the minimum tree and maximum shrub height, which may vary between 5 and 8m, approximately.
*Forest fallow:* this type stands for all complexes of woody vegetation deriving from the clearing of forest land for shifting agriculture. It consists of a mosaic of various reconstitution phases and includes patches of uncleared forests and agriculture fields that cannot be realistically segregated and accounted for area-wise, especially from satellite imagery. It excludes areas where site degradation is so severe that a reconstitution of the forest is not possible. Such areas should be included under 'shrubs' or outside woody vegetation.

*Source:* FAO, 1988

per cent) of the world's 'closed forests' (Table 2.1). The remaining two-thirds of the total land area was divided nearly equally between uses for agriculture, including pastures and 'other uses'. The latter ranged from deserts and glaciers to roads, mines and urban conglomerates. This broad picture helps one appreciate the relative importance of forest areas in the global ecosystem.

**Table 2.1** *Distribution of the World's Forest Lands (areas in millions of hectares)*

| | | Total forest and wooded lands | | Closed forest | | | Other wooded areas | |
|---|---|---|---|---|---|---|---|---|
| Region | Total land area | Area | % of total land area | Area | % of forest and wooded land | Total | Open | Fallow |
| Temperate | 6417 | 2153 | 34 | 1590 | 74 | 563 | na | na |
| North America | 1835 | 734 | 40 | 459 | 63 | 275 | na | na |
| Europe | 472 | 181 | 38 | 145 | 80 | 35 | na | na |
| USSR | 2227 | 930 | 42 | 792 | 85 | 138 | na | na |
| Other countries | 1883 | 309 | 16 | 194 | 62 | 115 | na | na |
| Tropical | 4815 | 2346 | 49 | 1202 | 25 | 1144 | 734 | 410 |
| Africa | 2190 | 869 | 40 | 217 | 25 | 652 | 486 | 166 |
| Asia and Pacific | 945 | 410 | 43 | 306 | 10 | 104 | 31 | 73 |
| Latin America | 1680 | 1067 | 64 | 679 | 63 | 388 | 217 | 170 |
| World | 13,077 | 4499 | 34 | 2792 | 62 | 1707 | 734 | 410 |

na = not available

*Source:* World Resources Institute (WRI), 1988 (based mostly on FAO data)

**Table 2.2** *Preliminary Estimates of Tropical Forest Area and Rate of Deforestation for 87 Countries in the Tropical Region*

| *Sub-region* | *Number of countries studied* | *Total land area* | *Forest area 1980* | *Forest area 1990* | *Area deforested annually 1981–1990* | *Rate of change 1981–1990 (% per annum)* |
|---|---|---|---|---|---|---|
| | | | *(thousands of hectares)* | | | |
| **Latin America:** | 32 | 1,675,600 | 922,900 | 839,900 | 8400 | -0.9 |
| Central America and Mexico | 7 | 245,300 | 77,000 | 63,500 | 1400 | -1.8 |
| Caribbean sub-region | 18 | 69,500 | 48,800 | 47,100 | 200 | -0.4 |
| Tropical South America | 7 | 1,360,800 | 797,100 | 729,300 | 6800 | -0.8 |
| **Asia:** | 15 | 896,600 | 310,800 | 274,800 | 3500 | -1.2 |
| South Asia | 6 | 445,600 | 70,600 | 66,200 | 400 | -0.6 |
| Continental South-east Asia | 5 | 192,900 | 83,200 | 69,700 | 1300 | -1.6 |
| Insular South-east Asia | 4 | 258,100 | 157,000 | 138,900 | 1800 | -1.2 |
| **Africa:** | 40 | 2,243,300 | 650,400 | 600,100 | 5100 | -0.8 |
| West Sahelian Africa | 8 | 528,000 | 41,900 | 38,000 | 400 | -0.9 |
| East Sahelian Africa | 6 | 489,600 | 92,300 | 85,300 | 700 | -0.8 |
| West Africa | 8 | 203,200 | 55,200 | 43,400 | 1200 | -2.1 |
| Central Africa | 7 | 406,400 | 230,100 | 215,400 | 1500 | -0.6 |
| Tropical Southern Africa | 10 | 557,900 | 217,700 | 206,300 | 1100 | -0.5 |
| Insular Africa | 1 | 58,200 | 13,200 | 11,700 | 200 | -1.2 |
| **Total** | **87** | **4,815,500** | **1,884,100** | **1,714,800** | **17,000** | **-0.9** |

*Source:* FAO, 1991

Keeping in mind that a little over 70 per cent of the globe's surface consists of oceans, tropical forests occupied about 5 per cent of the total.

Deforestation in tropical regions reached alarming rates in the late-20th century. The FAO estimated that, on balance, between 1980 and 1990 about 17 million hectares of forests in 87 countries in tropical regions had been converted to other land uses each year. This represented an annual deforestation rate in these countries of 0.9 per cent (Table 2.2). Presumably, expanding areas of crops and pastures had replaced a large portion of the forests that had disappeared.

At first glance, global land use data do not seem to lend much support to this hypothesis. World food production increased by 25 per cent between 1983 and 1993, but the area in arable land and permanent crops expanded by only 1 per cent (FAO, 1995). These global aggregates, however, hide a number of different processes. In many places good agricultural lands were being appropriated for urban, infrastructural or industrial uses and degraded crop lands were being abandoned, while other lands were being brought into farms, often at the expense of forests. It is not possible to deduce the extent that tropical forests have been displaced by cropland and pasture from these global land use estimates.

This assumption that agricultural expansion is the principal culprit behind tropical deforestation is reinforced by numerous anecdotal observations together with a few more systematic studies. It is also consistent with the widely held Malthusian notion that agricultural area expands in tandem with (in linear proportion to) population growth. The world population in 1650 was estimated to have been about 0.5 billion and in 1700 about 0.65 billion people (Meadows et al, 1972), while in 1980 it was 4.4 billion, an increase of 680 per cent in 280 years. Meanwhile, the area estimated to have been in crops increased by 466 per cent (Table 2.3), although most spectacular increases seem to have taken place in North America. Cropland is estimated to have increased from 2 per cent of the Earth's land area in 1700 to a little over 11 per cent in 1980. Assuming an average increase in crop yields of about 45 per cent, this was proportional to the increase in population.

From these data, the relationship between population growth and crop area expansion seems to have been rather close at the

**Table 2.3** *Net Conversion of Land to Crops by Region, 1700–1980*

| | Area (million hectares) | | |
|---|---|---|---|
| *World region* | *1700* | *1980* | *% increase* |
| Tropical Africa | 44 | 222 | 405 |
| North Africa/Middle East | 20 | 107 | 435 |
| North America | 3 | 203 | 6667 |
| Central and South America | 7 | 142 | 1929 |
| South and East Asia | 86 | 399 | 464 |
| Former Soviet Union | 33 | 233 | 606 |
| Europe (except FSU) | 67 | 137 | 105 |
| Australia/New Zealand | 5 | 58 | 1060 |
| **Total** | **265** | **1501** | **466** |

*Source:* Roberts, 1996, p504.

global level. The wide differences between rates of cropland increase by regions, however, suggest more complex processes. These regional differences possibly could be explained by differential impacts of trade and modern technologies. One suspects, however, that in the absence of firm data about either populations or cultivated areas in much of the world in 1700, the apparently close relationship between the growth of these two variables may have been influenced by the assumptions they are purported to demonstrate.

## RECENT LAND USE CHANGES

The more one delves into the data on land use changes at national levels in developing countries, the more qualifications one has to make concerning the usefulness of any simplistic generalizations. Obviously, if land use is divided into only three categories, 'forests', 'crops and pastures' and 'other uses', then each category includes a great many widely divergent processes. These range from intensively managed forest plantations to severe forest degradation, high-yielding cropping to extensive grazing, and urban development to desertification.

Many institutions and scholars believe agricultural expansion to be the major factor contributing to deforestation in developing countries. The FAO estimates that by the early 1980s, 70 per cent of the disappearance of closed forests in Africa, 50 per cent in Asia and 35 per cent in Latin America was due to the conversion of forest land to agriculture (FAO, 1982). The World Bank asserts that (during the 1980s) new settlements for agriculture accounted for 60 per cent of tropical deforestation (World Bank, 1992, p20). Myers, based on a survey of 28 tropical countries, concludes that by the late 1980s, agricultural expansion (exclusive of cattle ranching) was responsible for over two-thirds of the area that was deforested (Myers, 1989, p2). The WRI and WWF have also stated the permanent conversion of forest to agricultural land was the principal cause of deforestation in developing countries (WRI, 1990, pp106–107; WWF, 1989, pp9–11). There are many other NGOs, research organizations and specialists that have postulated varying rates of deforestation and the role of agricultural expansion to tropical forest clearance. However, most of these appraisals are based mainly on FAO estimates, supplemented by limited case study materials and differing interpretations.

In any case, even if agricultural expansion did account for half or more of the loss of tropical forest area in some regions in recent years, this would tell one little about why this was occurring or what could be done to halt it. The social dynamics of deforestation are much more complex than is usually admitted by conservationists and many other concerned observers (Barraclough and Ghimire, 1995).

The FAO has played a leading role in compiling information on changes in land use in major developing regions and countries. Data are available for some countries from the 1950s onwards, but it is only after 1970 that the details are provided for most countries in a more consistent manner. In recent years, the FAO has published estimates of total land use in each country broken down by arable land, permanent crops, permanent pasture, forest and woodland and other land uses. These data suggest a declining trend in forest coverage in most developing countries, but agricultural expansion does not seem to account for much of this deforestation at regional or national levels.

The changes in land use in developing countries at the regional level in Africa, Latin America and Asia between 1977 and 1992 are presented in Table 2.4. No reliable comparable data are available for earlier periods, in part because of the frequent changes in the estimated land areas of these continents. Table 2.4 suggests that during the period considered, agricultural land in

**Table 2.4** *Land Use Changes in Africa, Asia and Latin America (1977–1992)*

| | 1977 | 1982 | 1987 | 1992 | Difference between 1977–1992 (1000 ha) | (%) |
|---|---|---|---|---|---|---|
| **Africa** | | | | | | |
| Total land | 2,996,075 | 2,996,075 | 2,996,075 | 2,996,075 | – | – |
| Land area | 2,930,454 | 2,930,454 | 2,930,454 | 2,930,454 | – | – |
| Arable–permanent crops | 169,293 | 172,065 | 177,183 | 181,878 | 12,585 | 7 |
| Permanent pasture | 884,876 | 881,395 | 887,454 | 892,210 | 7,334 | 1 |
| Forest–woodland | 719,583 | 706,632 | 692,825 | 678,105 | -41,478 | -6 |
| Other land | 1,156,702 | 1,170,362 | 1,172,992 | 1,178,212 | 21,510 | 2 |
| **Latin America** | | | | | | |
| Total land | 2,051,257 | 2,051,257 | 2,051,257 | 2,051,257 | – | – |
| Land area | 2,015,444 | 2,015,444 | 2,015,444 | 2,015,444 | – | – |
| Arable–permanent crops | 131,436 | 140,399 | 146,587 | 150,988 | 19,552 | 15 |
| Permanent pasture | 558,418 | 569,321 | 581,608 | 590,481 | 32,063 | 6 |
| Forest–woodland | 962,162 | 935,746 | 908,339 | 880,782 | -81,380 | -8 |
| Other land | 363,428 | 370,038 | 378,910 | 393,193 | 29,765 | 8 |
| **Asia/Oceania** | | | | | | |
| Total land | 2,749,626 | 2,749,626 | 2,749,626 | 2,749,626 | – | – |
| Land area | 2,669,674 | 2,669,674 | 2,669,674 | 2,669,674 | – | – |
| Arable–permanent crops | 444,094 | 447,066 | 452,210 | 455,142 | 11,048 | 2 |
| Permanent pasture | 673,773 | 706,756 | 766,009 | 798,374 | 124,601 | 18 |
| Forest–woodland | 590,231 | 570,341 | 553,228 | 552,503 | -37,728 | -6 |
| Other land | 961,576 | 945,453 | 898,184 | 863,590 | -97,986 | -10 |

*Note:* Includes 110 selected countries from African, American and Asian continents. Countries with a 'land area' less than 500,000 hectares and industrialized countries were not included while Greenland and Djibouti were excluded due to their lack of 'arable land', and Qatar, Oman, Lesotho and the Falkland Islands were excluded due to their lack of 'forest land'.

*Source: FAO Production Yearbook, 1993*

Africa, Latin America and Asia increased by 8, 21 and 20 per cent respectively. Similarly, forest areas decreased by 6, 8 and 6 per cent respectively. In Africa and Latin America, forests appear to have been, in part, victims of agricultural expansion. Nearly half of the deforested areas in Africa and over one-third in Latin America, however, were accompanied by corresponding increases of areas with land uses other than for crops and pastures. Among these other uses were urbanization, infrastructure and industry as well as the abandonment of degraded barren lands. In Asia, on the other hand, much more land has been brought into agriculture than the areas that were deforested. It is clear that most of the new agricultural area in Asia came from 'other land' (for example, marginal areas). This is understandable as most of the agriculturally suitable land as well as easily accessible forest areas had already been exploited. In recent decades, however, this continent has also seen a rapid increase in settled areas, cities and the development of infrastructure. The data do not indicate where such 'developed' areas came from, but one suspects many of them had been used for agriculture.

Table 2.5 shows land use changes reported by the FAO between the 1950s and the early 1970s and 1992 for 110 developing countries.[1] It can readily be seen that land use trends differed greatly from one country to another. To bring out this diversity, the individual countries are grouped into seven categories in Table 2.6. Each group shows different trends in respect to changes in areas of agricultural land, forest land and other land.

At national levels the data indicate diverse trends in different groups of countries:

- In 14 countries agricultural area increased while the area of forests and woodlands and area of other land uses both decreased.
- In 33 countries agricultural area increased, the area of forests decreased and the area in other uses increased.
- Agricultural and forested areas both increased while other land decreased in 19 countries.
- Agricultural and forest areas both decreased while other land areas increased in 21 countries.
- Agricultural area decreased but both the forest area and the area in other uses increased in seven countries.

1 FAO's forest resources assessment for tropical countries provides estimates of forest areas, changes in forest areas 1981 to 1990, areas logged, forest ecological zones, forest formations and the annual deforestation rates for each, for 89 tropical countries (FAO, 1993). These data are undoubtedly more accurate concerning forest areas than those in Table 2.5 taken from the *FAO's Production Yearbook,* but they do not show changes in agricultural areas nor in land areas devoted to other uses. Given the focus of this book on agricultural expansion and tropical deforestation, the *Production Yearbook* data were used for consistency, but data from the forest assessment are cited in the text when appropriate in discussing country case studies. Two books analysing tropical deforestation issues have been published since the FAO's detailed forest assessment data became available (Brown and Pearce, 1994; Palo and Mery, 1996). These include many analyses and insights that will be referred to in later chapters.

**Table 2.5** *Land Use Changes in Developing Countries, 1950–1992*

| | *Agricultural land* | | *Forest land* | | *Other land* | |
|---|---|---|---|---|---|---|
| | *1000 ha* | *%* | *1000 ha* | *%* | *1000 ha* | *%* |
| Afghanistan (1973–92) | 6 | 0.02 | 0 | 0.0 | -6 | -0.02 |
| Algeria (1957–92) | -8605 | -18.2 | 970 | 31.6 | 7635 | 4.1 |
| Angola (1953–92) | 2600 | 8.7 | 8700 | 20.1 | -11,300 | -21.9 |
| Argentina (1973–92) | -9050 | -5.1 | -1500 | -2.5 | 10,550 | 30.2 |
| Bahamas (1973–92) | 2 | 20.0 | 0 | 0.0 | -2 | -0.3 |
| Bangladesh (1973–92) | -72 | -0.7 | -339 | -15.2 | 411 | 38.3 |
| Belize (1973–92) | 21 | 25.0 | 0 | 0.0 | -21 | -1.8 |
| Benin (1964–92) | -1120 | -32.5 | 1243 | 57.6 | -323 | -5.7 |
| Bhutan (1973–92) | 39 | 10.6 | 80 | 3.2 | -119 | -6.4 |
| Bolivia (1973–92) | -1341 | -4.4 | -1950 | -3.4 | 3290 | 15.8 |
| Botswana (1973–92) | -10,506 | -23.5 | 9928 | 1032.0 | 578 | 5.2 |
| Brazil (1973–92) | 36,780 | 17.6 | -101,850 | -17.3 | 65,070 | 139.1 |
| Brunei Darsm (1973–92) | -5 | -27.8 | -205 | -50.0 | 210 | 212.1 |
| Burkina Faso* (1973–92) | 1190 | 9.6 | -1120 | -14.7 | -90 | -1.2 |
| Burundi (1967–92) | 639 | 39.1 | -31 | -26.7 | -823 | -79.8 |
| Cambodia (1967–92) | 836 | 23.5 | -1772 | -13.3 | 484 | 41.4 |
| Cameroon (1973–92) | 880 | 6.1 | -2070 | -7.8 | 1190 | 21.0 |
| Central African Republic (1973–92) | 160 | 3.3 | -180 | -0.5 | 20 | 0.1 |
| Chad (1968–92) | -3744 | -7.2 | -3950 | -23.9 | 5165 | 8.6 |
| Chile (1965–92) | 3244 | 22.2 | -11,886 | -57.5 | 8642 | 21.8 |
| China (1973–92) | 75,846 | 18.0 | 16,871 | 14.8 | -92,717 | -23.3 |
| Colombia (1973–92) | 4715 | 11.4 | -6500 | -11.7 | 1785 | 25.1 |
| Congo (1963–92) | -4760 | -31.9 | 4870 | 30.0 | -160 | -5.3 |
| Costa Rica (1973–92) | 822 | 40.1 | -710 | -30.2 | -112 | -15.8 |
| Côte d'Ivoire (1968–92) | -149 | -0.9 | -4920 | -41.0 | 4623 | 136.5 |
| Cuba (1973–92) | 459 | 7.8 | -60 | -2.5 | -503 | -17.5 |
| Cyprus (1957–92) | -368 | -69.8 | -48 | -28.1 | 415 | 182.8 |
| Dominican Republic (1973–92) | 245 | 7.4 | -39 | -6.0 | -206 | -23.1 |
| Ecuador (1968–92) | 3157 | 65.8 | -4545 | -30.6 | 716 | 8.2 |
| Egypt (1973–92) | -255 | -8.9 | 0 | 0.0 | 255 | 0.3 |
| El Salvador (1973–92) | 79 | 6.3 | -70 | -40.2 | -22 | -3.4 |
| Equatorial Guinea (1963–92) | 9 | 2.8 | -994 | -43.4 | 985 | 515.7 |
| Ethiopia (1968–92) | -20,170 | -25.6 | 18,100 | 205.7 | -10,020 | -29.1 |
| Fiji (1973–92) | 142 | 48.5 | 0 | 0.0 | -142 | -40.7 |
| French Guyana (1973–92) | 14 | 200.0 | -300 | -3.9 | 286 | 23.7 |
| Gabon (1974–92) | 10 | 0.2 | -150 | -0.8 | 140 | 22.7 |
| Gambia (1967–92) | -330 | -55.0 | -158 | -52.1 | 358 | 157.7 |
| Ghana (1974–92) | -210 | -2.6 | -1260 | -13.7 | 1222 | 20.8 |

**Table 2.5** *Continued*

| | *Agricultural land* | | *Forest land* | | *Other land* | |
|---|---|---|---|---|---|---|
| | *1000 ha* | *%* | *1000 ha* | *%* | *1000 ha* | *%* |
| Guatemala (1950–92) | 1250 | 60.8 | -1242 | -25.7 | -54 | -1.3 |
| Guinea (1974–92) | -610 | -8.9 | -1080 | -6.9 | 1676 | 76.0 |
| Guinea-Bissau (1974–92) | 55 | 4.0 | 0 | 0.0 | -55 | -14.6 |
| Guyana (1973–92) | 350 | 25.4 | -1821 | -10.0 | 1485 | 1414.2 |
| Haiti (1950–92) | 535 | 61.5 | -665 | -95 | 111 | 9.2 |
| Honduras (1955–92) | 1438 | 48.0 | -1716 | -35.6 | 258 | 7.6 |
| India (1973–92) | 1810 | 1.0 | 3070 | 4.7 | -4880 | -9.3 |
| Indonesia (1973–92) | 3040 | 9.7 | -15,627 | -12.8 | 10,587 | 38.3 |
| Iran (1960–92) | 43,836 | 239.1 | 6020 | 50.2 | -51,056 | -38 |
| Iraq (1973–92) | 280 | 3.1 | -60 | -3.1 | -220 | -0.7 |
| Israel (1973–92) | 34 | 6.3 | 13 | 11.5 | -47 | -3.3 |
| Jamaica (1973–92) | 1 | 0.2 | -18 | -8.9 | 17 | 4.2 |
| Jordan (1973–92) | 86 | 7.7 | 12 | 20.7 | -98 | -1.3 |
| Kenya (1974–92) | 240 | 0.6 | -360 | -13.5 | 120 | 0.9 |
| Korea (DPR) (1973–92) | -80 | -3.7 | 0 | 0.0 | 80 | 8.7 |
| Korean (Rep) (1973–92) | -109 | -4.8 | -164 | -2.5 | 273 | 28.0 |
| Kuwait (1973–92) | 7 | 5.2 | 0 | 0.0 | -7 | -0.4 |
| Laos (1973–92) | -37 | -2.3 | -1900 | -13.2 | 1937 | 27.5 |
| Lebanon (1968–92) | -10 | -3.1 | -15 | -15.8 | 8 | 1.3 |
| Liberia (1977–92) | 4 | 0.06 | -400 | -19.04 | 396 | 26.3 |
| Libya (1959–92) | 4403 | 39.8 | 238 | 51.5 | -4641 | -2.8 |
| Madagascar (1954–92) | -1165 | -3.0 | 3450 | 28.8 | -3131 | -35.9 |
| Malawi (1959–92) | 17 | 0.5 | 1096 | 47.4 | -3553 | -59.1 |
| Malaysia (1973–92) | 301 | 6.5 | -3588 | -15.6 | 3287 | 61.9 |
| Mali (1974–92) | 403 | 1.3 | -530 | -7.1 | 127 | 0.2 |
| Mauritania (1974–92) | 38 | 0.1 | -180 | -3.9 | 142 | 0.2 |
| Mexico (1973–92) | 1170 | 1.2 | -11,210 | -21.5 | 10,040 | 24.7 |
| Mongolia (1973–92) | -14,482 | -10.3 | -1085 | -7.2 | 15,567 | 1609.8 |
| Morocco (1966–92) | 15,198 | 97.7 | 2541 | 47.4 | -17,764 | -74.8 |
| Mozambique (1974–92) | 100 | 0.2 | -2150 | -13.3 | 2050 | 13.5 |
| Myanmar* (1973–92) | 76 | 0.7 | 215 | 0.7 | -333 | -1.4 |
| Namibia (1974-92) | 9 | 0.02 | -600 | -3.2 | 591 | 2.4 |
| Nepal (1973–92) | 461 | 11.8 | 3010 | 128.6 | -3471 | -46.6 |
| Nicaragua (1963–92) | 4980 | 277.8 | -3250 | -50.4 | -2855 | -60 |
| Niger (1974–92) | -368 | -2.9 | -1060 | -35.8 | 1428 | 1.3 |
| Nigeria (1961–92) | 24,790 | 52.1 | -20,292 | -64.2 | -5798 | -44.0 |
| North Caledonia (1973–92) | -35 | -13.3 | 0 | 0.0 | 35 | 4.1 |
| Pakistan (1973–92) | 1729 | 7.1 | 1197 | 42.0 | -2926 | -5.9 |
| Panama (1977–92) | 301 | 16.3 | -1060 | -24.9 | 759 | 56.9 |
| Papua New Guinea (1973–92) | 47 | 10.5 | -330 | -0.9 | 283 | 4.5 |
| Paraguay (1954–92) | 22,748 | 1861.5 | -7150 | -35.8 | -16,543 | -85 |
| Peru (1966–92) | 895 | 3.0 | -19,000 | -21.8 | 17,583 | 152.0 |

**Table 2.5** *Continued*

| | *Agricultural land* | | *Forest land* | | *Other land* | |
|---|---|---|---|---|---|---|
| | *1000 ha* | *%* | *1000 ha* | *%* | *1000 ha* | *%* |
| Philippines (1967–92) | 1079 | 11.5 | -4603 | -31.5 | 3341 | 55.6 |
| Puerto Rico (1968–92) | -108 | -19.1 | 50 | 39.4 | 54 | 27.4 |
| Rwanda (1963–92) | -245 | -13.1 | 394 | 252.6 | -316 | -51.5 |
| Saudi Arabia (1973–92) | 37,716 | 43.8 | 199 | 12.4 | -37,915 | -29.8 |
| Senegal (1973–92) | -5149 | -48.6 | 4100 | 64.6 | 1049 | 45.5 |
| Sierra Leone (1964–92) | -3124 | -53.2 | 1739 | 577.7 | 1373 | 136.6 |
| Solomon Islands (1973–92) | 7 | 7.9 | 0 | 0.0 | -7 | -4.7 |
| Somalia (1960–92) | 22,513 | 104.6 | -5361 | -37.2 | -18,184 | -65.3 |
| South Africa (1960–92) | -7891 | -7.7 | 410 | 10.0 | 7481 | 48.1 |
| Sri Lanka (1973–92) | -11 | -0.5 | 300 | 16.7 | -289 | -12.5 |
| Sudan (1973–92) | 54,975 | 80.9 | -7720 | -14.9 | -47,255 | -40.2 |
| Surinam (1966–92) | 37 | 71.2 | -32 | -0.2 | -732 | -50.7 |
| Swaziland (1967–92) | -261 | -17.1 | -11 | -8.5 | 256 | 301.2 |
| Syria (1968–92) | 2659 | 23.5 | 215 | 48.9 | -3014 | -44.5 |
| Tanzania (1973–92) | -1454 | -3.6 | -3375 | -7.7 | 4584 | 100.1 |
| Thailand (1973–92) | 4850 | 30.2 | -6510 | -32.5 | 1660 | 11.1 |
| Togo (1974–92) | -746 | -23.3 | -311 | -17.7 | 1057 | 223.0 |
| Trinidad & Tobago (1957–92) | -8 | -5.7 | -57 | -20.7 | 65 | 67.0 |
| Tunisia (1973–92) | 1502 | 20.2 | 156 | 31.8 | -1658 | -21.7 |
| Turkey (1973–92) | 1056 | 2.7 | 29 | 0.1 | -1085 | -6.0 |
| Uganda (1973–92) | -1610 | -15.8 | -808 | -12.8 | 2482 | 70.0 |
| United Arab Emirates (1973–92) | 26 | 12.2 | 1 | 50.0 | -27 | -0.3 |
| Uruguay (1973–92) | -232 | -1.5 | 56 | 9.1 | 176 | 9.7 |
| Vanuatu (1973–92) | 49 | 40.8 | 898 | 5612.5 | -947 | -87.4 |
| Venezuela (1973–92) | 1509 | 7.5 | -5505 | -15.7 | 3996 | 12.1 |
| Viet Nam (1973–92) | 628 | 9.8 | -4000 | -29.3 | 3172 | 25.0 |
| Yemen (1973–92) | 39 | 0.2 | -1270 | -38.8 | 1231 | 3.8 |
| Zaire* (1973–92) | 500 | 2.2 | -6110 | -3.4 | 5586 | 22.9 |
| Zambia (1973–92) | -4707 | -11.8 | -1790 | -5.9 | 6497 | 167.9 |
| Zimbabwe* (1973–92) | 378 | 5.2 | -930 | -4.7 | 570 | 5.0 |

*Notes:* The same 110 developing countries that were included in Table 1.4.

Agricultural land includes arable land and permanent crops plus permanent pasture.

* Means that in these countries the total area had changed between 1973 and 1992: Burkina Faso decreased by 20,000 hectares, Zaire decreased by 24,000 hectares, Zimbabwe increased by 18,000 hectares and Myanmar increased by 3000 hectares.

*Source: FAO Production Yearbooks, 1958–1961, 1969–70, 1989–90 and 1993.*

**Table 2.6** *Diverse Trends of Agricultural Expansion and Deforestation in 110 Developing Countries, 1950–1992*

| *Trend* | *Country* | *Years* |
|---|---|---|
| *Increase in agricultural land. Decrease in forest. Decrease in other land.* | Burkina Faso | 1973–92 |
| | Burundi | 1967–92 |
| | Nigeria | 1961–92 |
| | Somalia | 1960–92 |
| | Sudan | 1973–92 |
| | Costa Rica | 1973–92 |
| | Cuba | 1973–92 |
| | Dominican Republic | 1973–92 |
| | El Salvador | 1973–92 |
| | Guatemala | 1950–92 |
| | Nicaragua | 1963–92 |
| | Paraguay | 1954–92 |
| | Surinam[b] | 1966–92 |
| | Iraq | 1973–92 |
| *Increase in agricultural land. Decrease in forest. Increase in other land.* | Cameroon | 1973–92 |
| | Central African Republic[b] | 1973–92 |
| | Equatorial Guinea | 1963–92 |
| | Gabon[a, b] | 1974–92 |
| | Kenya[a] | 1974–92 |
| | Liberia[a] | 1977–92 |
| | Mali[a] | 1974–92 |
| | Mauritania[a] | 1974–92 |
| | Mozambique[a] | 1974–92 |
| | Namibia[a] | 1974–92 |
| | Zaire | 1973–92 |
| | Zimbabwe | 1973–92 |
| | Brazil | 1973–92 |
| | Colombia | 1973–92 |
| | Chile | 1965–92 |
| | Ecuador | 1968–92 |
| | French Guyana | 1973–92 |
| | Guyana | 1973–92 |
| | Haiti | 1950–92 |
| | Honduras | 1955–92 |
| | Jamaica[a] | 1973–92 |
| | Mexico[a] | 1973–92 |
| | Panama | 1977–92 |
| | Peru | 1966–92 |

**Table 2.6** *Continued*

| | | |
|---|---|---|
| | Venezuela | 1973–92 |
| | Cambodia | 1967–92 |
| | Indonesia | 1973–92 |
| | Malaysia | 1973–92 |
| | Philippines | 1967–92 |
| | Thailand | 1973–92 |
| | Viet Nam | 1973–92 |
| | Yemen [a] | 1973–92 |
| | Papua New Guinea | 1973–92 |
| *Increase in agricultural land. Increase in forest. Decrease in other land.* | Angola | 1953–92 |
| | Libya | 1959–92 |
| | Malawi[a] | 1959–92 |
| | Morocco | 1966–92 |
| | Tunisia | 1973–92 |
| | Bhutan | 1973–92 |
| | China | 1973–92 |
| | India[a] | 1973–92 |
| | Iran | 1960–92 |
| | Israel | 1973–92 |
| | Jordan | 1973–92 |
| | Myanmar [a, b] | 1973–92 |
| | Nepal | 1973–92 |
| | Pakistan | 1973–92 |
| | Saudi Arabia | 1973–92 |
| | Syria | 1968–92 |
| | Turkey [b] | 1973–92 |
| | United Arab Emirates | 1973–92 |
| | Vanuatu | 1973–92 |
| *Decrease in agricultural land. Decrease in forest. Increase in other land.* | Côte d'Ivoire[a] | 1968–92 |
| | Chad | 1968–92 |
| | Gambia | 1967–92 |
| | Ghana | 1974–92 |
| | Guinea | 1974–92 |
| | Niger | 1974–92 |
| | Swaziland | 1967–92 |
| | Tanzania | 1973–92 |
| | Togo | 1974–92 |
| | Uganda | 1973–92 |
| | Zambia | 1973–92 |

**Table 2.6** *Continued*

| | | |
|---|---|---|
| | Argentina | 1973–92 |
| | Bolivia | 1973–92 |
| | Trinidad & Tobago | 1957–92 |
| | Bangladesh [a] | 1973–92 |
| | Brunei Darsm | 1973–92 |
| | Cyprus | 1957–92 |
| | Korean Republic | 1973–92 |
| | Laos | 1973–92 |
| | Lebanon | 1968–92 |
| | Mongolia | 1973–92 |
| *Decrease in agricultural land. Increase in forest. Increase in other land.* | Algeria | 1957–92 |
| | Botswana | 1973–92 |
| | Senegal | 1973–92 |
| | Sierra Leone | 1964–92 |
| | South Africa | 1960–92 |
| | Puerto Rico | 1968–92 |
| | Uruguay [a] | 1973–92 |
| *Increase/Decrease in agricultural land. No variation in forest. Increase/decrease in other land.* | Egypt | 1973–92 |
| | Guinea-Bissau | 1974–92 |
| | Bahamas | 1973–92 |
| | Belize | 1973–92 |
| | Afghanistan[a] | 1973–92 |
| | Korea (DPR) | 1973–92 |
| | Kuwait | 1973–92 |
| | Fiji | 1973–92 |
| | New Caledonia | 1973–92 |
| | Solomon Islands | 1973–92 |
| *Decrease in agricultural land. Increase in forest. Decrease in other land.* | Benin | 1964–92 |
| | Congo | 1963–92 |
| | Ethiopia | 1968–92 |
| | Madagascar | 1954–92 |
| | Rwanda | 1963–92 |
| | Sri Lanka[a] | 1973–92 |

a A variation of less than 2 per cent in 'Agricultural land'
b A less than 2 per cent variation in 'Forest land'
*Source:* Table 1.5

- In 10 countries, forest area remained stable, accompanied by increases and decreases in agricultural area and other land uses.
- In six countries agricultural and other land uses both decreased while forest area expanded.

Agricultural expansion appears to have been a significant factor in explaining deforestation in some countries but not in others. Clearly, detailed analyses at national and sub-national levels are required in order to understand better the social dynamics of agricultural expansion and tropical deforestation.

# 3 Tropical Deforestation and Agricultural Expansion in the Case Study Countries

In four of the five case study countries (the exception was China) there had been a significant decrease reported in forested areas between 1973 and 1992.[1] Table 1.1 suggested that in Brazil, forest area had decreased by 17 per cent, accompanied by an increase in cropland and pasture equal to 36 per cent of the deforested area and in other lands equal to 64 per cent of the missing forests. In Guatemala, forest area decreased by 28 per cent, with 67 per cent of this loss accounted for by an increase in other lands and 33 per cent by expansion of agricultural areas (crops and pastures). The loss of forest area in Cameroon was 8 per cent, with 57 per cent of the lost forest area going to 'other land' and 43 per cent to agricultural uses. Lost forest areas in Malaysia during these two decades accounted for 16 per cent of the 1973 forest area with almost all of it (92 per cent) accounted for by an increase in 'other land' and 8 per cent by agricultural expansion. In China, forest area apparently expanded by 15 per cent while the areas under pasture and crops grew by 18 per cent, with the increased areas in both agriculture and forests accompanied by an equivalent diminution of 'other land'.

These national level land use estimates suggest that in the case study countries losses in forest area were for the most part to 'other land', with a much smaller portion being cleared for agricultural expansion. On the other hand, as was seen in Chapter 1, this is contradicted by the findings of numerous studies. This chapter examines the evidence from the case study countries and especially

1 The data in Table 1.1 are from the FAO's *Production Yearbooks* that compile their estimates on the basis of reports by governments. Hence, they differ somewhat, and tend to be less accurate and comparable, than are the regional data reported in Table 2.2 based on the FAO's inventory of tropical forests. For the purpose of indicating gross national level land use trends, these differences are of minor significance.

that from sub-national data and local level studies. The case study information is analysed to the further extent possible with reference to the six clusters of issues set forth in Chapter 1.[2]

## PUBLIC POLICY-INDUCED DEFORESTATION IN BRAZIL[3]

Brazil includes the largest area of tropical forest currently found in any individual country. Of a total land area of 846 million hectares, over half (488 million hectares) was estimated to have been forested in 1992. About five-sixths of this forest area was in the country's Amazon region. Nearly three-fourths of these Amazonian forests were classified as moist tropical forests (tropical rainforests). Deforestation in the Amazonian region did not become significant, however, until the mid-20th century. The Brazilian Institute for Space Research (INPE) estimates that between 1975 and 1991 some 30 million hectares of Brazil's Amazonian forests were cleared for other land uses. This was equivalent to an annual deforestation rate of a little over 2 million hectares per year. About one-tenth of the country's Amazonian forest area reported in 1950 had been cleared by 1991, with most of this loss occurring after 1975.

While there had been little forest clearance in the Amazonian region before 1950, there had been a great deal of earlier defor-

2 The case study summaries that follow may appear overly descriptive to readers conditioned to associate analyses with correlation matrices and quantitative models. One should recall that analysis always implies the description of relationships among a system's component parts at differing levels of generality, whether these relationships are described in symbolic or literary terms. We believe that comparative analyses of deforestation processes in widely differing and frequently changing socioeconomic, political and ecological contexts can be better communicated for most readers through descriptions of the interactions, associations and linkages observed than through falsely precise statistical or mathematical formulations. The case studies here attempt to describe the relationships encountered at different levels among deforestation processes, institutions, the policies of diverse social actors, livelihoods and the natural environment. If done competently, this is analysis in the true sense of the concept. These narratives tell us that the roles of agricultural expansion and trade in tropical deforestation cannot be understood, or feasible remedies suggested, in specific contexts without relating them to other components of the systems in which these processes are taking place.

3 The material used in this section is largely based on the case study reports cited below and summarized in English in Angelo-Furland and de Arruda Sampaio, 1995, and Barraclough and Ghimire, 1995.

estation in the coastal region south of the Amazon. In the 16th century the Europeans had found dense sub-tropical forests of over 100 million hectares extending from Brazil's north-east (south of its Amazon basin) to what is now the frontier with Uruguay and Argentina. Over 90 per cent of these original Atlantic coastal forests (the Mata Atlantica) had been cleared by the mid-20th century. They had been converted to croplands, pastures, urban space, roads and other infrastructure, to wastelands and also in some places to forest plantations. This earlier massive deforestation had commenced with the expansion of sugar-cane plantations in Brazil's north-east that were worked mostly by slaves brought from Africa from the beginning of the 16th century. The rapid expansion much later of other export crops, such as coffee and cocoa in the 19th and early 20th centuries, had contributed to accelerated clearance of the remaining Atlantic coastal forests.

As in most of Latin America, the expansion of commodity production for export had been one of the principal processes driving Brazil's economic and demographic growth since the European conquest. Sugar exports in the 16th and 17th centuries had stimulated massive deforestation associated with forest clearance for sugar production, as well as for feeding and housing the influx of European colonists and African slaves associated with the expansion of this lucrative export crop. Sugar cane expansion slowed in the 18th and 19th centuries but agricultural commodity exports accelerated again with the expansion of coffee, cocoa and rubber in the 19th century. The rapid expansion of coffee production, in particular after 1870, stimulated widespread deforestation. As slavery was legally abolished in Brazil in 1888, workers had to be induced by other means to produce these non-traditional export crops. Agro-export expansion after the mid-19th century was accompanied by large scale immigration from Europe, the Near East and, to a much lesser extent, from Japan. This immigration was actively encouraged and often highly subsidized by the government and by large private exporters. In this sense, increasing population in much of rural Brazil was more a result of, than a cause of, agricultural expansion. Also, for the most part, production of these newer export crops, such as sugar early on, was controlled by a few large landowners.

Brazil's export-led development strategy changed to one of greater emphasis on import substitution and industrialization when faced with the constraints on exports and imports accompanying the great depression and the Second World War. Import substitution and industrialization continued to be a high priority in subsequent decades although it was complemented by a boom in soybean exports after the 1950s. The foreign debt crisis of the 1980s, and several other factors, induced the state to renew efforts to attract foreign investments in a more open economy. Recent deforestation processes have, of course, been influenced by such changes in the state's dominant development strategy.

By 1990 Brazil had become an upper-middle-income country according to the World Bank's classification, with a per capita average national income similar to the average for all of Latin America. This was about the same as for Malaysia and three to five times higher than that of the other three case study countries. Moreover, it was by far the most urbanized, with only one-fifth of its workforce engaged in agriculture in the mid-1990s. Its agricultural workforce had actually decreased in absolute numbers after 1960, while average productivity per agricultural worker had more than doubled. Manufacturing had become far more important than agriculture in its contribution to GDP, while its export dependence on primary commodities had decreased from 90 per cent in 1960 to about 50 per cent in 1990. In spite of these impressive economic changes, the number and proportion of its people living in poverty had increased during the 1980s and the worst poverty was still to be found among its rural population.

Beginning in the 1940s, and especially after the military coup of 1964, the Brazilian state embarked on a massive campaign to 'occupy' the 'empty' Amazon region that included over half of the country. Huge state and private investments were undertaken. These included all-weather roads from Brasilia to Belém in Pará, from Cuiabá in Mato Grosso to Pôrto Velho in Rondônia, from Pôrto Velho to Manaus, from Manaus west through the Amazon forests towards the Andes, and many others. There were also gigantic investments made in hydroelectric projects, mining, industries, eucalyptus plantations, cattle-ranching and agro-export initiatives. Private investors received lucrative tax exemptions and other state subsidies. Colonists and workers were encouraged to leave the poverty-stricken north-east and areas in the south in order to find new livelihoods in the Amazon.

Land tenure rights were chaotic, with countless conflicts between land speculators, big investors and poor settlers as well as with long-time residents such as indigenous peoples, riverine peasants and rubber-tappers. The *latifundia* system that had dominated rural Brazil since the European conquest, enabling a few large landowners to control most rural resources and labour, was being extended to Amazonia in what some observers called the biggest land enclosure movement in all history (Hecht and Cockburn, 1990).

This brief sketch of national level trends provides the context in which the agricultural expansion and deforestation reviewed below in the sub-national level case studies occurred. Four of these case studies were in the 'legal Amazon' region, where most recent deforestation has taken place. One of the studies was in an area of south-eastern São Paulo state that includes part of the remaining 3 per cent of the original Mata Atlantica and where further deforestation is still a threat.

### *São Felix do Araguaia in north-eastern Mato Grosso*[4]

North-eastern Mato Grosso includes several million hectares, most of which were forested until recent decades. The municipality of São Felix on Mato Grosso's eastern border with the state of Tocantins covers over half a million hectares. The vast north-eastern Araguaia region was very sparsely populated until the mid-20th century. Most of it was occupied by indigenous tribes that included several distinct linguistic groups. Before recent settlement of the area by Brazilian immigrants, many of these indigenous peoples had had little contact with Brazil's Portuguese-speaking inhabitants. Ecologically the area is a transition zone between moist tropical rainforests and drier savannah. Dense tropical 'gallery' forests dominated the richer moist soils of river valleys, while more open woods, brush and grass covered drier sites.

Deforestation in north-eastern Mato Grosso was particularly rapid during the 1970s and early 1980s. It apparently slowed

4 Based on A Umbelino de Oliveira, 1995

afterwards, primarily because there was little accessible forest left to clear in the region. Satellite data analysed by the INPE indicated that for the state of Mato Grosso as a whole, the deforested area had increased from nearly a million hectares in 1975 to 2.6 million hectares in 1978, to 6.7 million hectares in 1988 and 8.4 million hectares in 1990. A major part of this deforestation had taken place in the north-east region that included São Felix de Araguaia.

Deforestation in the São Felix area commenced on a very small scale in the 1940s when the government encouraged the migration of settlers from Brazil's impoverished north-eastern states and from Minas Gerais to seek improved livelihoods in Mato Grosso. They initially cleared small plots for self-provisioning, causing very limited ecological damage. As the forested areas were very extensive, this first small-scale slash-and-burn peasant agriculture did not generate many direct conflicts with indigenous groups, although it did increase their exposure to new infectious diseases.

In the 1960s the government offered attractive tax incentives and other subsidies for large-scale investors in the Amazon region. This induced land speculators, big ranchers and agro-industrialists to seek control of the best lands in north-eastern Mato Grosso. Specialists in the fabrication of land titles (*grileiros*) abounded. As the areas occupied by indigenous peoples were considered state lands, these lands were rapidly claimed by outsiders, often in huge estates of tens of thousands of hectares. This occurred in spite of constitutional provisions prohibiting the sale of state lands in lots of more than 10,000 hectares by the 1946 constitution, 3000 hectares in the 1967 constitution and 2500 hectares in that of 1988. These limits were easily circumvented by individuals or corporations obtaining titles to adjacent properties under fraudulent names.

This occupation of the region by land speculators was supplemented by a massive state-sponsored colonization programme through the national colonization and land reform institute (INCRA) in the 1970s. The land was first cleared of its indigenous inhabitants and of peasant squatters who had been encouraged to settle earlier. It was then sold to private developers who undertook to prepare it for agricultural settlement and subdivide it into 'family' farms for sale to settlers. These projects were often corrupt in execution as well as in the land acquisition. They seldom made available the technical or other support to the colonists that they had been paid to provide.

State lands for settlement projects were sold to corporate, individual and cooperative 'developers' at very low prices (usually for the equivalent of $3–$7 per hectare) who in turn, after removing timber with commercial value, sold lots to settlers. The new owners cleared and burned forests for crops and pasture. The soils soon lost their fertility and markets were uncertain. Crops inevitably failed and debts could not be paid. The settlers then sold their shares in the land at very low prices or simply abandoned it for work in new nearby towns. The land quickly became concentrated in the hands of a few large cattle growers, speculators and other big owners. It was usually sown to pasture but it could not support more than about one head of cattle per hectare. Within four or five years, soil fertility had declined rapidly, often requiring two hectares per head of cattle. Labour requirements were very low as only one worker was required to care for up to about 2000 cattle. Within a few years much of the once forested land was virtually abandoned.

Much the same happened with the millions of hectares in the case study area that were never 'colonized' but directly taken over by large ranchers or agro-industrial corporations. The steps of subdivisions and forest clearing by settlers were replaced by inducing immigrant workers or share-croppers under inhumanely harsh conditions to clear the forests and brush to make way for pasture. Once lavish government subsidies were reduced in the mid-1980s, many of these areas also became unproductive for their owners and were often virtually abandoned or sold to speculators. Some areas of better soil and water resources, however, were farmed under the management of large estates for cash crops such as soya and rice.

Several bloody social conflicts were engendered by these processes. Most new immigrants faced highly exploitative labour conditions. A 'gold rush' to northern Mato Grosso during the same period stimulated additional conflicts. Many thousands of indigenous inhabitants died from disease or were killed by thugs hired by large landowners. Squatters ousted from the lands they occupied fared little better. Moreover, most of the immigrant workers and peasant colonists encountered unbearable conditions that eventually drove them to the cities or to new areas in the Amazon that were being deforested.

### *Bico do Papagaio in Tocantins*[5]

This case study included six municipalities with an area of about one and a half million hectares in the state of Tocantins, lying north-east of the Mato Grosso study summarized above. As in the São Felix case, the vegetation is transitional between moist tropical forest and savannah. A greater portion of the soils, however, tend on the whole to be apt for intensive agricultural uses than in north-eastern Mato Grosso. Also, settlement by Portuguese-speaking Brazilians began nearly two centuries earlier here than in Mato Grosso. Early settlers included run-away slaves who arrived from the north-east in the 18th century to practise slash-and-burn agriculture and the collection of forest products such as *babaçu* (an oil-yielding palm nut). There were many conflicts with indigenous inhabitants who had mostly disappeared by the 20th century as a result of flight, extermination and absorption.

The population increased very slowly from in-migration and natural growth until the mid-20th century. Peasant farmers were able to use rotations of from five to seven years in areas of slash-and-burn that were sufficient to maintain fertility. Cash incomes came from the sale of some rice, beans, manioc and the like when production exceeded subsistence needs, and from the collection and sale of *babaçu*. In some upland areas it came from cotton that was mostly produced by share-croppers on traditional large estates. Most of the land was used in smallholdings without clear boundaries as their sizes were continuously adapted to changing family labour forces and other needs. The land tenure system was based on custom but was apparently effective in defining the rights and responsibilities of individual land users as well as those of local communities.

The situation began to change in the 1950s with increased immigration. It changed very rapidly in the 1960s and 1970s when the policies of occupying the Amazon described earlier were

5 Based on Regina Sader, 1995.

introduced by the military government. By 1950, nearly two centuries after the earliest colonization in the case study region, and six decades after a few forested areas began to be converted to cattle ranches and other commercial agricultural uses, the total population in the six municipalities was about 16,000 people, of whom only 3400 were living in towns. Deforestation was restricted to small areas of slash-and-burn agriculture, a few riverine flood plains and some pastures in relatively sparsely forested uplands. The completion of the Belém–Brasilia highway in the 1960s provided access to the region for poor immigrants, big corporate enterprises and land speculators. By 1970 the population had increased to over 130,000, with over 50,000 in towns and most of these in the rapidly growing city of Imperatriz. By 1991 the region's total population had reached some 450,000, of whom over 100,000 lived in Imperatriz and some 45,000 in other towns.

Deforestation has been spectacularly rapid. INPE satellite information shows that in 1973 closed forests covered half the region, forests interspersed with clearings another 30 per cent, while the remaining 20 per cent of land was in concentrated clearings. Eleven years later 80 per cent of the region was in concentrated clearings. By 1992 only 6 per cent of the area remained in forests, 4 per cent in forests interspersed with clearings and 90 per cent was in concentrated clearings. Comparing a soil map and a land capabilities map for the region with the land use maps, one readily sees that many of the areas cleared for pasture and crops were ill adapted for these uses on a sustainable basis.

The processes resulting in this massive deforestation were almost identical to those described above in the São Felix case. Land speculators rapidly fabricated titles with the cooperation of some state officials and agencies. The rights of earlier occupants were ignored. State subsidies in the form of tax favours, cheap credits (often at negative real interest rates), huge subsidized investments and the like made acquiring land titles by whatever means, and then deforesting the land, a highly lucrative business. Among several other large projects, the state subsidized the construction of 22 saw-mills, of which only half were actually constructed.

Many bloody conflicts ensued between small farmers and the agents of speculators and large foreign and national corporate investors. The military government embarked on a programme of land privatization that reached its height in the 1970s. In 1970, farm units of less than 100 hectares each were reported to have accounted for 81 per cent of the region's cultivated area (excluding pastures), while they constituted 74 per cent of the number of all farms. Twenty years later in 1990, units of less than 100 hectares constituted 88 per cent of all farms but included only 17 per cent of the cultivated area. Peasant farming systems had been largely replaced by industrial large-scale monocropping and by extensive cattle ranches. This is an almost inevitable outcome where markets are little influenced by the rural poor, but instead are largely controlled by big producers and their allies.

### *Pillage of the Riktbaktsa in north-eastern Mato Grosso*[6]

Until the mid-20th century, the Riktbaktsa people occupied a territory of some 5 million hectares in north-western Mato Grosso. Most of this area was tropical rainforest. Dense rainforest dominated non-inundated areas in the north, open tropical forests in the central zone and deciduous forests in the south. Soils have little potential for agriculture in the south but about half have limited possibilities for crops or pasture in the centre and north of the region. Very little of the region's soils have a potential to support 'modern' agriculture or ranching. They are well adapted, however, for the Riktbaktsa people's hunting, gathering and itinerant farming, essential for their social reproduction.

The Riktbaktsas numbered only about 1200 in the 1950s, living in some three dozen small settlements. Their numbers had already been reduced by diseases communicated by contact with rubber tappers, prospectors and missionaries and possibly by brief contact early in the century with workers constructing a telegraph line from Cuiabá (capital of Mato Grosso) to the frontier state of Acre in the north-west of Brazil. The first conflicts reported between the indigenous peoples and settlers occurred in the 1950s, with the advent of more intensive rubber tapping and a few private colonization projects.

In 1961 the government created the Juruena Forest Reservation that included most of the Riktbaktsa's territory and promised the

6 Based on RSV Arruda, 1995.

Riktbaktsa use of the entire area. This forest reserve never became operational. After the 1964 military coup the government created a reserve for the Riktbaktsa, confining them to only 10 per cent of their territory. In practice this opened up the remainder to land speculation, timber exploitation, settlement and mining activities without restrictions, while protection of the smaller indigenous reserve itself was problematic.

By 1969 disease, social disruption and violence had reduced the Riktbaktsa's numbers to less than 300. Confined to their small reservation, missionaries (Missão Anchieta – MIA) brought them improved health services that helped to increase their numbers to about 700 by the late 1980s. The missionaries, however, contributed to the disruption of traditional Riktbaktsa society, rendering them dependent on missionary handouts and on cash income from wages and the sale of handicrafts.

Most of the Riktbaktsa territory was appropriated for commercial use during the 1980s. A principal instrument in this pillage of their traditional lands was the 'Polonoroeste' programme. This US$1.5 billion programme was 30 per cent financed by the World Bank. The highway through Riktbaktsa territory connecting Cuiabá with Pôrto Velho was improved and paved, making the region accessible to land speculators and lumber companies. Within two years of the paving of the road, lumber enterprises had devastated 2 million hectares. Land speculators quickly sold titles to lands in Indian territory. The Brazilian government's National Indian Foundation (FUNAI) imposed a development model that promoted settled agriculture and extensive cattle ranching, complemented by large-scale commercial rice and soybean production, mining and lumbering. This resulted in widespread forest clearance and the concentration of land ownership in a few large landowners.

The means used to expel the Indians from their former lands were similar to those described in the earlier two case studies. Roads were built to open up the territory for commercial exploitation. There were no prior government measures, institutions or agencies to control who occupied the land or to protect the rights of the original occupants. Squatters were encouraged to clear subsistence plots and, by implication, to displace the traditional indigenous inhabitants. Timber enterprises began to exploit the forest. Land speculators (*grileiros*) violently expelled the squatters and the remnants of the indigenous occupants. Social conflicts followed, ensuring that government agencies had to intervene to 'restore law and order'. Indigenous peoples were confined to small reservations while squatters and other traditional residents were forced to migrate to other areas. According to the case study report, the government then recognized fraudulent land titles held by very large enterprises and individual owners and protected their interests by using its police and judicial powers (Arruda, 1995). Development agencies granted subsidized credits to production and colonization projects of these big enterprises that had been approved by the state's bureaucracy.

Satellite images in 1992 showed nearly one-fourth of 634,000 hectares immediately adjacent to the reduced Riktbaktsa reserves had been converted to pasture (85 per cent) and large-scale commercial crops such as rice and soybeans (15 per cent). Most of it had been cleared since 1985. This does not reveal the extensive deforestation due to lumbering where the land had not yet been cleared for other uses nor does it show the prevalence of smaller clearings interspersed in remaining forest areas. In contrast to the surrounding area, there had been almost no forest clearance within the remaining half a million hectares of the Riktbaktsa reserve area except for a few places where land grabbers or miners had illegally pursued their activities. How long protection of the reduced reserve areas will endure, however, remains an open question.

### *The Mirassolzinho settlement and deforestation in south-western Mato Grosso*[7]

Mirassolzinho is a squatter settlement in Jaruru municipality. This is a small municipality by Mato Grosso standards, covering only some 200,000 hectares. It had been heavily forested until the 1960s but is now mostly in big and medium-sized properties largely cleared of forests to make way for pastures and some large-scale commercial monocultures. This region lies within the 'legal Amazon' but just south of the Amazon basin. Unlike the Riktbaktsa case summarized earlier that lies about 700 kilometres

7 Based on BAC Castro Oliveira, 1995.

to the north within the Amazonian drainage system, Jaruru is in the drainage system of the Paraguay river near the divide between the two river basins.

As south-eastern Mato Grosso is on the frontier with Bolivia, it was considered a strategic region first by the Portuguese and later by the Brazilian state. Hence, it was penetrated by occasional small military expeditions and garrisons since the 17th century as well as by prospectors seeking gold and precious stones. In the 1920s there was also some movement into south-western Mato Grosso by cattle ranchers from the south-east, but these were of minor consequence for the forests as they mainly occupied savannah areas south of Jaruru. Forest clearance in the Jaruru area did not commence until the 1950s and did not become important until the 1970s and 1980s, when it was stimulated by the military government's programme to occupy the Amazon region. As with the São Felix do Araguaia area to the north-east, deforestation was rapid. In the 1970s it was promoted by subsidies for 'development projects'. It accelerated in the 1980s with the completion of the Cuiabá–Pôrto Velho highway and was also stimulated by several other components of the Polonoroeste programme. A gold rush in nearby areas brought more immigrants in the early 1980s. The surviving indigenous population had been confined to reservations after many conflicts in the 1960s.

Poor immigrants from Brazil's east, especially from Minas Gerais, were attracted to Jaruru in the 1960s by land speculators, ranchers and timber enterprises who claimed the land and who allowed the immigrants to cultivate parcels for self-provisioning while cutting the virgin forest for timber at less than subsistence wages, subsequently to make way for pastures and commercial crops. Once valuable timber was removed and the remaining forests cleared, most of these migrants had to move on to repeat the process in other still forested regions.

About 200 peasants in the study area, however, had continued to squat on a 2000 hectare holding of an absentee land speculator living in São Paulo instead of moving on. In the early 1980s this owner's heirs hired *pistoleros* (armed thugs) to remove the squatters. The peasant occupants resisted with machetes and shotguns, waging a virtual guerrilla war for several years with casualties on both sides. With the return of a civilian government nationally, the squatters were promised that the land they occupied would be expropriated by INCRA and redistributed to its occupants. In 1993 they still lacked legal titles and feared that speculators would again attempt to oust them in order to capitalize on rising land prices. Meanwhile, they had established a viable community and had adopted peasant farming systems that were far more sustainable ecologically and socially than the large ranches and commercial farms that dominated most of the surrounding area.

Actually, land concentration in Jaruru was not as extreme as in most of Mato Grosso or in much of the rest of Brazil. The 1985 agricultural census showed 32 properties of over 1000 hectares each and one of these was 27,000 hectares. This 2.5 per cent of the farm units included 60 per cent of the land. At the same time nearly 83 per cent of the farms were less than 100 hectares each and included 16 per cent of the land in farms. Most larger owners claimed 'legal' titles while most smallholders were squatters or tenants and hence not eligible to receive official subsidized credit. Land titles had commonly been initially acquired by *grileiros*, as in the other case study areas.

The case study looked particularly at the squatter community's social organization and its farming practices. It highlights the environmentally friendly nature of this peasant farming system in contrast to that of large commercial cattle ranches and big plantations of commercial crops such as soybeans and sugarcane. The peasant farmers initially grew corn, rice, beans, manioc and the like for self-provisioning. Their fields and pastures were established on better soils near streams or other sources of water interspersed with woodlands on poorer sites. Farm units were commonly five to ten hectares in size depending largely on the family labour force. In the 1980s they marketed increasing shares of their production in response both to growing demand in the region and to increasing farm productivity. Large ranches, in contrast, faced declining soil fertility once the soil nutrients left from the forest were depleted. A typical 3000 hectare ranch could maintain only from 2000 to 2500 cattle, and it provided year-round employment to about two full-time workers, although part-time workers were employed seasonally to clear brush from pastures.

In the late 1980s some government agencies and NGOs were able to obtain access to state and World Bank funds that became available in Mato Grosso to assist small farmers. This was in part a political response to the widely publicized ecological and social disasters associated with the original Polonoroeste project. The technical assistance and credits provided to small producers, however, were frequently wasted, largely because of the lack of real participation by the intended beneficiaries. In Jaruru, a small-farmer cooperative was in practice controlled by state and NGO employees. Large NGO-channelled investments designed to produce charcoal from *babaçu* nuts and sugar from cane for local use failed. They were soon swallowed by the jungle. Another cooperative was forced to assume large debts for vehicles that were later 'stolen'. A huge million-dollar state investment in an automated factory to convert manioc and maize into flour packaged for commercial sale was abandoned, leaving small producers with more unpayable debts.

INCRA's project to provide titles to the squatters was resisted by its intended beneficiaries, who had never been consulted. Instead of respecting the informal land divisions already made on the basis of ecological conditions and family labour supplies, INCRA proposed to give each family rectangular 25-hectare lots laid out on a map in its urban office with no regard for variable topography, ecology or family needs.

### *The evolution of deforestation in the Ribeira de Iguape river basin*[8]

The Ribeira de Iguape river basin is located in the south-eastern corner of the state of São Paulo, with a small portion extending south-westwards into the state of Paraná. The São Paulo part of the basin covers 1.7 million hectares. This is about 7 per cent of the total area of the state of São Paulo, but in 1990 it included some 60 per cent of the entire state's remaining forest area. In 1990, the state as a whole was only 7 per cent forested, but the Ribeira basin was still 54 per cent under forest cover. In the 16th century, however, 82 per cent of the state was estimated to have been forested, while forests had covered 95 per cent of the Ribeira basin. The Ribeira basin forests now include the largest remaining remnants of Brazil's once extensive Atlantic coastal forests.

8 Based on S Angelo-Furland and F A de Arruda Sampaio, 1995.

The Ribeira river drops steeply from the São Paulo plateau (at an elevation of about 1000 metres) to sea level near Iguape, only some 100 kilometres to the east. Rainfall is abundant, averaging nearly 2 metres annually, and the climate is sub-tropical. The region's geologic history together with heavy rainfall and the sharp drop in elevation produced an extremely rugged topography. Most soils are poorly suited for modern agriculture except in flood plains and small scattered valleys. These ecological constraints largely explain the Ribeira basin's slow deforestation compared to that in the rest of the state. Only some 3 per cent of the Ribeira basin area had been deforested by 1962. In São Paulo state, excluding the Ribeira basin, on the other hand, almost 17 million hectares of forest (over 90 per cent of its forested area) had been cleared between 1854 and 1962.

Deforestation in the area accelerated sharply after the 1950s. Over 40 per cent of the Ribeira basin's forest cover was eliminated between 1952 and 1985, while much of the remainder was badly degraded. Soils and topography in the rest of the state were mostly well suited to commercial farming, making it profitable to clear the forests to produce cash crops such as coffee, sugarcane, cattle and other commercial activities as soon as domestic and export markets became available after the mid-19th century. Rapid deforestation after 1950 in the Ribeira basin requires other explanations than agricultural expansion in spite of its proximity to rapidly industrializing metropolitan São Paulo.

The Ribeira basin's indigenous population (the Guaina) had caused little deforestation. Sixteenth- and 17th-century Portuguese colonists were drawn to the area principally in search of gold. Gold production became important during the 17th and 18th centuries. This resulted in a significant population increase requiring food. The 19th-century imperial government vigorously promoted agriculture in the region and it became an exporter of rice and manioc to other parts of Brazil. A canal built in the mid-19th century to make Iguape a deep-water port soon silted, slowing cash crop expansion. In any event, agricultural activities had caused very limited deforestation in the region. The arrival of Japanese colonists in the early 20th century stimulated rice production and also introduced the cultivation of tea and several other commercial crops such as tomatoes. Banana production in the valley for the São Paulo urban market increased rapidly after the 1920s. Also, there had been some commercial logging to meet demands from urban São Paulo. Nonetheless, by the early 1950s only 41,000 hectares of the region's original forest of over 1.5 million hectares in 1500 had been converted to other uses (2.8 per cent of the original forest cover).

Small peasant farming communities producing mostly for their own self-provisioning had been established in scattered locations throughout the river basin during the colonial and imperial periods. For example, slaves abandoned by their owners in the 17th century when gold was exhausted in the middle valley established the community of Ivaporunduva. Its church was constructed in 1630 and 78 families, all descendants of the original slaves, continue to farm the area. They developed sustainable farming systems that require five- to ten-year fallows with second growth forest vegetation for many of their crops. These peasants have no legal titles to their lands, however, even though their forebears had been farming it for three centuries. Moreover, new environmental legislation has made them criminals if they continue with their traditional farming and extractive practices. Also, a hydroelectric project threatens to flood their lands and homes, but as squatters they may not receive compensation.

Many similar peasant communities were established later in various localities throughout the region, especially after the 1920s. These peasant farming systems, however, have been responsible for only a minute fraction of the deforestation that occurred after 1950.

In the 1950s urban São Paulo had already become Brazil's wealthiest and largest city. A new highway through the Ribeira basin linking São Paulo with Paraná contributed to a speculative rise in land values. Many of the region's previously inaccessible forests became attractive for commercial logging. Also, land became potentially valuable for recreational uses as well as for the possible development of hydroelectric projects and more intensive uses of other natural resources. Under the military dictatorship after 1964, the region was briefly a site of guerrilla activity. Secondary roads were constructed by the army, making land speculation even more attractive.

The military government imposed no restrictions on speculators (*grileiros*) and absentee owners acquiring titles to large holdings. Low or non-existent land taxes, liberal tax credits for 'land improvement', cheap credits for large landowners and land values consistently increasing more rapidly than inflation, made land speculation profitable for those who could afford the initial investment. As soon as valuable timber was sold, most large landowners converted large portions of their forest areas to pasture. This was not because cattle-growing in the area was economically valuable but rather because forest clearance was legally considered to constitute 'land improvement'. In a situation where overlapping and conflicting land titles were common, owners had big incentives to replace forests with pastures in order to legalize their property and to obtain state subsidies even if the pastures were of dubious and declining productivity.

As mentioned above, most peasant farmers in the region have no valid legal titles to the lands they occupy even though they and their ancestors may have held it for generations. Many more recent settlers were given provisional titles upon making an initial payment, but when their lands were claimed by speculators the authorities frequently found the provisional titles illegal even though the small landholders had receipts for their purchase and had been paying annual taxes for many years. Thousands of peasants have lost their lands, without compensation, to speculators and to government development or conservation projects. They have had to move on to other forest areas, to city slums or to remain as illegal and precarious squatters.

In 1990, half of the land holdings were of less than 10 hectares each, but included only 3.7 per cent of the land. Large owners of over 100 hectares controlled 70 per cent of the land but accounted for only 8 per cent of the number of farm units reported. Most small units were operated by squatters or tenants.

By the end of the 1980s, nearly 60 per cent of the region's lands had been legally incorporated into parks or other strictly protected areas. This has left thousands of peasant families as illegal squatters on what many believed to be their own lands. Conflicts have been widespread, as have been the loss of traditional livelihoods by rural people. Meanwhile, land speculation for recreational development has become a major contributor to deforestation in several areas.

The rural population in the region during the period of rapid deforestation actually decreased from 169,000 in 1970 to 154,000 in 1980. In 1970, nearly 70 per cent of the region's total population was rural but in 1980 only 40 per cent were rural residents. By 1991 the rural population had increased slightly to a little over its 1970 level, but the urban population accounted for nearly two-thirds of the region's 485,000 inhabitants. The Ribeira basin was still by far the least densely populated region in the state of São Paulo.

Deforestation slowed markedly after 1985, in part due to economic recession and in part to conservation initiatives by NGOs and the state as well as the extensive deforestation that had already taken place in accessible areas potentially valuable for other land uses. Recent conservation efforts, however, have been uncoordinated. Their negative implications for the rural poor were not adequately taken into account, generating a host of new social and ecological issues.

### *Conclusion*[9]

The five local level case studies just reviewed support the hypothesis that government policies have been the principal cause of recent tropical deforestation in Brazil. Rural population pressures and careless peasant farming practices have at best played only a very minor role. Land tenure and related institutions, however, have been major factors both in generating the policies that encouraged undesirable deforestation and in determining their negative social and ecological impacts.

State policies aimed at integrating 'unoccupied' regions into the national economy have been particularly decisive in accelerating deforestation. Alternative approaches could have been taken with much lower social, ecological and economic costs and with much greater potential benefits. These alternatives, however, would have required the mobilization and support of popularly based social forces. Land settlement policies contributed to undesirable tropical deforestation as they were designed to be a substitute for,

9 Based on Plinio Arruda Sampaio and Francisco Arruda Sampaio, in Angelo-Furland and de Arruda Sampaio, 1995.

instead of a complement to, badly needed agrarian reform. Moreover, both economic integration and settlement policies were accompanied by widespread corruption of all kinds.

Policies ostensibly designed to protect the environment as well as those to protect indigenous populations and other groups of the rural poor have been ineffective in approaching their stated objectives. On the contrary, they have frequently contributed to the negative social and environmental impacts associated with undesirable tropical deforestation. The inefficacy of government environmental and social protection policies may often be a purposeful 'non-policy' designed to produce irreversible situations on the ground that contribute to short-term profits by powerful state support groups. On balance, they accelerated tropical deforestation. Several aspects of the Brazilian and the other country cases will be considered again in concluding chapters.

## DEFORESTATION AND AGRICULTURAL EXPANSION IN GUATEMALA[10]

Guatemala is a small central American country with an exceptionally rich biodiversity. Mountain ranges extending from the Atlantic coast in the east to the Pacific in the west divide the country into a relatively dry southern zone and a humid region in the north, while sharp differences in altitude ranging from sea-level to over 4000 metres have resulted in numerous distinct ecosystems in each region. The country was almost entirely forested when the Spaniards arrived in 1500, although there was some transitional forest-savannah vegetation in the south. Forests now cover about 40 per cent of the country and they are disappearing rapidly.

A dense indigenous population in what is now the Petén in the north had constituted part of the Maya civilization that flourished in Meso America from the 6th century BC to the 10th century AD. Mayan city-states were abandoned some five or six centuries before the arrival of the Europeans. Deforestation of much of the Petén to make way for maize and bean production to feed

10 Based largely on I Valenzuela, 1996.

Mayan urban centres, as well as to supply them with timber and fuel, is believed by many authorities to have contributed to this precipitous decline. The Petén was again heavily forested in the 16th century.

Present-day Guatemala consists of 109,000 square kilometres, 2 per cent the size of Brazil. Its population of 10.3 million in 1994, however, was 6.5 per cent that of Brazil's, making Guatemala rather densely populated by Latin American standards. Moreover, its population continues to be predominantly rural (about 60 per cent in 1994) in spite of urban population growth being twice as rapid as in rural areas during recent decades. Total population was increasing during the 1980s at an annual rate of nearly 3 per cent.

Guatemala's population is predominantly indigenous, especially in most rural areas. In all the Americas, only Bolivia has a similarly high proportion of its people who retain their indigenous languages and many of their traditional customs. Some two-thirds of Guatemala's rural population speak one of over two dozen native dialects as their mother-tongue and a big proportion of them have little knowledge of Spanish. From early colonial times the indigenous peoples were considered to be 'wards' of the conquerors, with few autonomous rights. Their forced labour for large land owners and 'public works' was legally sanctioned until the Arevelo reforms of the late 1940s. They were the principal victims of the civil war that resulted in over 150,000 deaths during the last three decades. For five centuries Guatemala's indigenous peoples have been dominated by a mainly Spanish-speaking oligarchy that controlled most political, economic and cultural institutions including the lands and other natural resources it deemed to be commercially valuable. The country's social structure has become increasingly complex, especially since the growth of modern commercial agro-exports in the 20th century. Nonetheless, the widely perceived divide between 'whites' (*criollos* of European descent) and *ladinos* (of mixed European and indigenous descent), on the one hand, and indigenous peoples, on the other, is a social reality that has to be taken into account in dealing with deforestation issues.

Closely related to this ethnic cleavage is the control of agricultural land by a small landowning élite. Land ownership in

Guatemala is probably the most concentrated of any country. Some 10,000 large landowning families and corporations controlled over two-thirds of all agricultural land in the late 1970s and this situation continues nearly unchanged. There are estimated to be over 800,000 rural households, most of whom are nearly landless or landless. Of 600,000 farm units recorded in the 1979 census, 60 per cent were of less than 1.4 hectares each. In other words, a little over 1 per cent of the agricultural population controls two-thirds of the land while 70 per cent are landless or nearly landless. The majority of these rural families with inadequate and insecure access to sufficient land to enable them to have acceptable livelihoods are indigenous peoples. A great many of the rural poor and destitute, however, are descendants of European immigrants and consider themselves to be whites or *ladinos*. Cultural differences reinforced by a long history of ethnic discrimination often impede these rural poor from uniting with their indigenous neighbours in attempting to bring about a more equitable agrarian structure.

Increasing the production of agro-export crops controlled by a few large land owners was a dominant objective of state policies since early colonial times. The best accessible farm lands were reserved initially for the production of export crops such as indigo, cochineal, cocoa and sugar. Coffee production for export became important in the late 19th and early 20th centuries. This stimulated the immigration of German coffee producers as well as coffee production by several native Guatemalan large landholders. Banana production for export near both the Atlantic and Pacific coasts also commenced on a large scale, principally on huge land concessions granted to the United Fruit Company, a United States-based transnational enterprise. Within 50 years after the so-called 'liberal' reforms began in the late 19th century, over a million hectares held by the church and by indigenous groups were confiscated by the state and granted to large coffee producers alone. Much more good agricultural land was taken from smallholders and allocated to large producers of other agro-exports. Most of these expropriated lands were soon deforested, while the displaced peasants had to move to less accessible areas where they cleared forests to make way for subsistence crops.

Agro-export expansion financed the 'modernization' of large-scale commercial agriculture together with much of the infrastructure and economic activities servicing it. It also produced a small class of capitalist entrepreneurs and a growing middle class. Political and social institutions, however, remained rigidly authoritarian and quasi-feudal. The Second World War cut off Guatemala's German market for coffee exports and left the coffee-dependent sector of the oligarchy in disarray. A middle class-led urban-based popular movement gained momentum, leading to the fall of the Ubico dictatorship and the recall of Arevalo from exile to head an elected reformist government in 1944.

Important social reforms were commenced in the late 1940s. These culminated in an important agrarian reform in 1952 during the democratically elected Arbenz administration. About a million hectares from 1000 expropriated large estates were distributed to some 100,000 landless peasants between 1952 and 1954. The reform was carried out with little violence or corruption and peasant beneficiaries received credit and technical assistance. Food production actually increased. This land reform sparked a violent reaction by the landowning oligarchy and by foreign investors such as the United Fruit Company. The Arbenz régime was overthrown by a United States-backed military coup in 1954. All the expropriated lands were taken from peasant beneficiaries and returned to former large owners. The counter reform was followed by four decades of military repression and costly civil conflict. Hopefully, the 1996 peace settlement will help bring this ugly period to an end.

Meanwhile, deforestation and the modernization of export-oriented commercial agriculture accelerated. In the 1950s close to 70 per cent of Guatemala was still covered by forests. This implied that about one-fourth of the country's forests had been cleared for agriculture and other uses since the Spanish conquest, with most of this forest clearance taking place during the early 20th century. By 1990, however, only about 40 per cent was forested. More forest had been cleared during the last three decades than during the previous five centuries. Much of the remaining forest has been badly degraded by logging and in some cases by over-exploitation for fuelwood.

Agricultural expansion after the 1950s continued to be a principal factor directly leading to deforestation. Forests were cleared both to make way for agro-export crops and to accommodate self-provisioning peasants displaced by large export crop producers and other concomitants of agricultural modernization such as mechanization. The production of new non-traditional exports such as cotton and beef expanded rapidly, in the 1960s and 1970s, while sugar and coffee production also continued to increase. Forests in big holdings on the southern Pacific coast were rapidly cleared for the production of cotton, pasture and sugarcane. Thousands of peasant families migrated yearly to the cities and thousands more were pushed to remaining forested frontiers such as the Petén. The United States government, the World Bank, the Interamerican Development Bank and many other international sources provided the government with low-cost loans to 'modernize' its economy. Large-scale agro-export producers received over 80 per cent of all agricultural credit in Guatemala between 1956 and 1980.

The deforestation stimulated by agro-exports and agricultural modernization was complemented by the state's colonization programme. This was initiated primarily to ease peasant pressures for agrarian reform after the Arbenz administration's reform had been annulled by the military government. There were a few small land settlement projects in the south, but the government's big colonization effort was in the north and especially in the Petén. It received international financial and technical assistance for a massive land settlement programme under the umbrella of the Alliance for Progress. As will be seen later in the discussion of the two local-level case studies, the deforestation processes associated with this programme were very similar to those analysed earlier in the Brazilian Amazon region. Few peasants benefited, but hundred of thousands of hectares of forest were cleared.

The Guatemalan economy grew by over 5 per cent annually during the 1960s and 1970s. It slowed badly during the 1980s when terms of trade deteriorated and the internal war took a heavy toll, but it picked up again in the early 1990s. In 1994 Guatemala was classified by the World Bank as a lower-middle-income country with an estimated average per capita national income of US$1200. Its income distribution, however, was estimated by the Bank to be the most skewed in all Latin America with the exception of much higher income Brazil. Some 70 per cent of Guatemala's rural population was classified as poor in the early 1990s and over half of these poor were destitute. In many indigenous regions nearly everyone was destitute (with incomes below only one half the poverty line).

At local levels deforestation processes were diverse and their social dynamics complex. A case study of deforestation in the department of Totonicapan in the western highlands was carried out in the early 1990s for UNRISD (Utting, 1993). Some 30,000 indigenous families lived in the region and most were cultivating parcels of less than one hectare each. Their production was primarily for self-provisioning and local markets. The community's customary livelihood system depended partly on forest products for artisanal furniture making and their sale throughout the country. Its forests also provided wood for fuel, local construction needs, fertilizers, water supplies and a host of other uses. Large areas of communally held pine forests had been kept in sustainable high-level production to meet the community's growing population's many needs even though nearly all forests had been cleared in neighbouring regions with similar ecological constraints and population densities.

This traditional land management system in Totonicapan broke down in the 1980s, however, when outsiders invaded the community forests to steal valuable white pine bark for sale to provide tannin for leather manufacture and also to rob merchantable timber. In spite of having legal title to the land and the community's efforts to protect its forest resources, the indigenous land owners were powerless to stop bark strippers and other thieves who were sanctioned by the state's military and police. Indian forest guards who attempted to resist were denounced as subversive guerrilla sympathizers.

For the present research, two additional case studies were carried out. One was in the Petén and the other in La Sierra de las Minas. These are summarized below.

### *Forest destruction and protection in the Petén*

In spite of having been heavily deforested by the Mayas over one thousand years earlier, in the mid-20th century four-fifths of the Petén was covered by moist tropical forests together with a few small areas of pine forest. Most of the Petén is a well watered low-lying plain only about 200 metres above sea-level. Soils tend to be shallow, lying directly on bedrock. Many of the region's soils can be moderately productive for the sustainable production of crops or pastures if carefully used as components of agro-forestry systems, but for the most part they are unsuited for modern high-external-input agriculture. The Petén covers 36,000 square kilometres of northern Guatemala, which is a little over one-third of the area of the entire country. It had only about 3 per cent of the country's population in 1990. Population growth in the region has been very rapid since the 1950s when the Petén had only a few thousand inhabitants. The area still includes two-thirds of the remaining forest area in all Guatemala. Deforestation in the Petén currently amounts to over 50,000 hectares annually.

Selective logging of high quality mahogany and cedar for export commenced in the 19th century. This was limited to areas accessible by rivers that made the transport of logs to seaports economically feasible. Rapid deforestation did not commence until the 1960s, when the construction of roads financed by international credits opened the region to timber exploitation, speculators, settlers and cattle ranchers. Cattle grazing had previously been restricted to accessible semi-savannah areas.

Land speculation exploded in the Petén as soon as plans for the construction of a road connecting Flores in central Petén with Guatemala City became known. Ranchers, timber merchants, developers and speculators of all sorts, both foreign and domestic, rushed to secure titles to large parts of the Petén. Their task was made easier by the military government, many of whose members joined in the quest for land titles. Moreover, the administration of the Petén was entrusted to a state-controlled corporate body responsible for the region's promotion and development (FYDEP) created in 1959. With the spread of guerrilla activities in the Petén this state corporation was promptly transferred to the department of defence, making the Guatemalan army the sole government of the Petén from 1960 until 1987, when other state agencies were permitted to operate there more normally.

Deforestation in the Petén proceeded in a pattern similar to that already described in Mato Grosso. Once roads were opened, commercially accessible, valuable timber was removed. Large ranchers and speculators encouraged poor migrant workers and tenants to clear remaining forest to grow subsistence crops. These were followed by pastures as soon as initial soil fertility declined after a year or two. At the same time, other poor migrants from the south commenced slash-and-burn agriculture in forest areas still regarded as state lands or claimed by large landholders with disputed titles. Others came as part of the state-sponsored colonization programme during the 1960s with promises of provisional titles and official assistance. Much of the land designated for colonization by smallholders, however, was in fact given to army officers and politicians. Credit and technical assistance seldom arrived. As forest clearance was regarded by the authorities as proof of improved land use, which was required to obtain land titles, both big speculators and poor migrants had an additional incentive to clear as much forest as possible.

The situation in the Petén was further complicated by guerrilla activities in the 1960s and 1970s. The army destroyed many forests along roads with chemical defoliants to reduce cover for guerrillas. Villages and cooperatives were often completely destroyed, with numerous residents killed. Many indigenous communities fled to Mexico. Lawlessness and violence were augmented by booming drug trafficking often controlled by the army itself. Poppy cultivation spread in many formerly forested areas.

As noted earlier, immigration to the Petén from the rest of rural Guatemala accelerated rapidly after 1960. Poor *ladino* peasants and farm workers arriving from the south-west and south-east found their agricultural practices poorly adapted for the Petén's soils and climate. Modern inputs were unavailable or unaffordable. Their soils quickly degraded and many abandoned their lands to cattle ranchers while they sought employment in towns or elsewhere. Indigenous immigrants usually brought more sustainable farming systems that included agro-forestry. Also, early *ladino* colonists readily adopted sustainable practices developed by other indigenous groups that had been long-term residents in the region. Peasant immigrants, however, usually had insecure rights to the lands they occupied, regardless of their origins and farming practices.

The rapid destruction of the Petén's forests has apparently not been slowed significantly by numerous initiatives ostensibly designed to protect them, nor have laws and agencies supposed to protect the rural poor been effective in doing so. The Petén includes many potential tourist attractions such as the ancient Mayan city of Tikal, a little north of Flores. It also has petroleum and many other natural resources in addition to its forests and agriculture. State agencies responsible for protecting these natural endowments and for their rational use, however, were virtually non-operative in the region from 1960 until 1987. When they were able to function normally, they were hampered by inadequate resources, poorly trained personnel, frequently corrupt administration and, above all, by conflicting political pressures and interferences. By the early 1990s numerous international NGOs and aid agencies were pouring millions of dollars into the Petén to help protect its flora and fauna. Many seemed oblivious to the socio-political obstacles faced on the ground by anyone attempting to foster more sustainable development of the Petén.

The Mayan Biosphere Reserve in the Petén that was established in the late 1980s covers nearly one-third of the entire area. It includes a strictly protected area of 800,000 hectares and a surrounding partially protected multiple-use zone of 650,000 hectares. Adjacent buffer zones were also created in which residents would benefit from special technical assistance and educational programmes. In 1994, however, the reserve was still subject to many invasions (especially by loggers) and conflicts. Some local officials feared loss of their prerogatives and authority. Peasants had never been consulted and feared loss of their livelihoods. The reserve was administered by the National Council for Protected Areas (CONAP). There was considerable bureaucratic competition between CONAP and other government agencies such as the National Forest Service (DIGEBOS) over their respective responsibilities and also their access to funds flowing in from foreign sources for environmental protection. Political support for the reserve's consolidation remains uncertain. Nonetheless, its creation represented a major initiative that constituted an official recognition of the seriousness of the region's environmental problems.

There were numerous other initiatives towards more sustainable development in the region, mostly by NGOs. The government was developing a National Forestry Plan of Action, but no practical effects had yet been seen in the Petén by the mid-1990s. The difficulties are enormous. For example, a journalist who exposed illegal smuggling of valuable timber to Mexico was attacked and beaten and forced into exile by death threats. Employees of the state agency that administers protected areas (CONAP) were attacked when they discovered timber robbers. NGO and government officials who reported massive timber smuggling into Belize to their respective head offices in Guatemala City were prohibited from returning to that area. Deforestation therefore seems likely to continue for some time still.

### *Deforestation and agriculture in La Sierra de las Minas*

In 1990 Guatemala's legislature approved a law creating the Biosphere Reserve of La Sierra de las Minas as a part of UNESCO's Man and the Biosphere (MAB) programme's world-wide network of nature reserves. The Reserve is planned to have a strictly protected nucleus of 105,000 hectares plus 4200 hectares of badly degraded land to be replanted, 34,600 hectares designated for regulated multiple uses, and 91,800 hectares of buffer zone between the strictly protected and multiple use area. It is administered by a council presided over by CONAP, but its direct administration is entrusted to a Guatemalan NGO, La Fundación Defensores de la Naturaleza (FDN). FDN is largely financed from the United States and other foreign sources. This Reserve is only about one-sixth the size of the Mayan Biosphere Reserve described above. Unlike the latter, which was entirely within the department of the Petén, the Sierra de la Minas Reserve cuts across 5 of Guatemala's 13 administrative departments.

La Sierra de las Minas is a mountain range extending some 120 kilometres north-east from the country's central high plateau north of Guatemala City down to its largest inland body of water, Lake Izabal, near the Atlantic Coast, just south of the frontier with Belize. The Reserve's width ranges between 10 and 30 kilometres and its total area is about 236,000 hectares. It is more or less bounded on the west by the main highway from Guatemala City to Cobán in the department of Alta Verapaz. Its northern boundary is the Polochic river valley, which is narrow and steep in its upper reaches but much broader and flatter near Lake Izabal. On the south it is bounded by the much wider and more densely populated Motagua valley. This valley also provides Guatemala City's rail and road connection with its Atlantic port of Puerto Tomás. The Motagua valley is one of the country's principal industrial zones as well as accounting for nearly one-fourth of Guatemala's irrigated land and nearly one-tenth of its total population. The two valley bottoms are excluded from the Reserve.

The Motagua valley and later the Polochic valley were mostly divided into large *haciendas* early in the colonial period, although a few indigenous communities remained in the Polochic area. Like the Petén, the Polochic valley had been a site of Mayan cities until they were abandoned some five centuries before Spanish colonization. La Sierra de la Minas took its name from mining activities that began on its steep slopes in the colonial period and some of which still continue.

Elevations in the Reserve range from mountain peaks of over 3000 metres to only some 200 metres above sea level near Lake Izabal. Rainfall reaches 4000 millimetres annually in the upper Polochic valley and exceeds 2000 millimetres along the Sierra's crest, but is only some 500 millimetres at lower elevations of its southern slopes. Soils tend to be rather shallow and subject to accelerated erosion when tree cover is removed. The Reserve has at least five major ecosystems as a result of big differences in elevation and rainfall. One of these ecological zones includes 60 thousand hectares of unique 'cloud forest' found only at elevations of over 1800 metres. These physical and climatic characteristics had generated an exceptional biodiversity in what is now the Reserve area.

The Reserve and its adjacent valleys had all been forested when the Spaniards arrived. The drier Motagua valley and slopes were covered by transitional savannah forest with extensive pine forests at higher elevations before giving way to the moist subtropical mixed forest cover that dominated most of the Reserve area's highlands and its northern slopes. Deforestation during the colonial period had mostly been limited to valley bottoms where trees were removed for construction, fuel and to clear land for pastures and crops. During the 19th and early 20th centuries there had been some deforestation within the Reserve area associated with mining and later on its southern slopes, with the construction of the railway from Guatemala City to Puerto

Tomás. Except for a few small peasant settlements on the southern slopes, there had been little forest clearance for agriculture in what is now the Reserve. This situation changed rapidly after the 1950s.

Modern commercial agriculture expanded rapidly in the Motagua valley after 1950, stimulated by irrigation, industrial development and improved export markets. Most good land in the valley floor was in large- and medium-sized holdings. The valley's rapidly growing population required increasing amounts of construction materials and fuelwood, as did nearby Guatemala City. In addition there was also a lucrative demand for good cedar, mahogany, oak and pine timber for export. As a result, commercial logging expanded rapidly up the Sierra's southern slopes. Logging activities also opened roads for a few peasant migrants and provided opportunities to pasture cattle and goats in previously forested areas. In a sub-tropical, low rainfall context, grazing together with seasonal burning to regenerate pasture resulted in severe desertification of the once forested lands above the valley floor up to about 1000 metres. This area is now mostly covered with desert-like thorny bushes and cactus. The land has become of such low value for pasture that it is not economical to construct new fences or to repair old ones. There are a few peasant farmers on the lower slopes settled near streams or springs where they can raise crops while obtaining supplementary low value pasture nearby for their livestock. Most settlements, however, are at higher elevations where rainfall is more abundant. In the early 1990s, there were estimated to be 27 small *ladino* settlements on the southern slopes of the reserve with a few dozen families in each. Some had been established in the late 19th century but others were of more recent origin.

The *ladino* settlement of Moran, for example, was first established a century ago at an elevation of about 1700 metres where a mountain stream provides water for residents and their cattle as well as a little for seasonal irrigation. The village is accessible from Rossario in the Motagua valley by a steep 12-kilometre track. It is transitable during the dry season. Older residents can remember when the area was heavily forested, moister and cooler. During the early 20th century some forest had been removed by slash-and-burn farming practices to cultivate maize and beans and a few areas had been converted to pasture. Massive deforestation did not take place around Moran, however, until the area was commercially logged in the early 1960s. Although the community supposedly had title to this forest area, it was never asked permission by the loggers and the peasants received no compensation.

Since the area was logged, peasants have faced growing water shortages and declining crop yields. The community now has 60 households but only 250 residents. These are disproportionately very young or well over middle-aged, as many members of working age have migrated or are working 'temporarily' elsewhere. Some have gone to the Petén or found employment in the valley, while a few went to the United States. The settlement survives partly from remittances sent back by its emigrant workers. As the peasants produce very little for sale outside their community, this income from remittances is crucial. Otherwise it would have been impossible to obtain the clothing, utensils, some foods and a few very modest conveniences such as medicines and an occasional beer or Coca-Cola that are now considered to be essentials. It also made possible the purchase of a little chemical fertilizer to maintain maize yields on exhausted soils, and zinc roofing for some of the huts in place of traditional thatch. The settlement also had a small one-room school for the first three grades, served by a teacher who hiked up from the valley on Mondays and to which he returned on Thursdays. Only about half the young children attended. Rudimentary school supplies were in short supply and there had been no 'school lunches' for several years.

In this settlement, agriculture was not expanding but contracting. It certainly had not been a major cause of recent deforestation. This was a rather typical situation in the buffer zone on the southern side of the Reserve. In the smaller settlement of Jones, a little further south, but accessible to Moran only via the Motagua valley, there had been even greater outmigration. There were also signs of greater 'prosperity', as indicated by two new four-wheel drive vehicles, a few television aerials, a small motor-driven electric generator and a few modern concrete cottages. There was little agriculture. Remittances from emigrants to the United States and connections of some residents with the drug trade through Guatemala to points north accounted for this relative affluence.

Higher up the mountains in the 'cloud forest' part of the Reserve's nucleus there still remain a few areas of magnificent moist deciduous forests. They had not been logged earlier because they had been commercially inaccessible. Logging had begun in the mid-1980s, however, and it had not yet been effectively stopped by the creation of a strictly protected area. A local small farmer dedicated to forest conservation, in part to protect the water supplies of his own adjacent farm, was employed as a forest guard by the FDN. He had been ambushed with his young son while marking the nucleus area boundary a little above his home. Thugs probably sent by the holder of a logging concession that might not be renewed fired on them with shotguns. He had been badly crippled, and his son eventually died from his wounds. Meanwhile, illicit logging continues. The FDN has no way of physically impeding it and the government authorities to which it can appeal apparently do not have either the means to prevent it or the necessary political support from key political and military figures, who some suspect may be profiting from the logging.

Deforestation on the northern slopes of the Reserve has also been serious, but more abundant rainfall facilitates natural forest regeneration after logging even when it is followed by slash-and-burn agriculture. The forest-friendly climate in the valley has stimulated commercial large-scale tree crops such as coffee, rubber and citrus plantations. As with the Motagua side, most serious forest degradation has taken place since the 1950s.

Traditional *haciendas* in Polochic valley had been dedicated principally to cattle raising. In the 1960s it became profitable for large land owners, most of whom lived in Guatemala City much or all of each year, to invest in the production of coffee, citrus, rubber and citronella (an Asiatic grass yielding oil valuable for perfumes and insect repellents). Some of these new export crops, however, required much more labour than traditional cattle ranching. The large land owners permitted the immigration of indigenous peasants from other parts of Alta Verapaz and elsewhere who were employed in the new plantations.

Plantation owners, however, usually prohibited these workers from using their lands to grow maize and beans for self-provisioning. This would not only have reduced the area available for commercial use but would also have tended to create a more independent workforce. Wages were much too low for the indigenous workers to support their families. As a result, workers clandestinely cultivated their subsistence crops on forested mountain slopes and frequently moved family members or brought relatives to tend them as squatters in the forests. These steep forest lands were mostly claimed by absentee large owners who had already sold the most valuable timber when it became possible to do so with the improvement of the valley's dirt road in the 1960s. Owners held the land speculatively after selling the timber, but they exercised little effective on-the-ground control over it. In this way large areas of 'highgraded' forest became used for slash-and-burn maize and bean production by indigenous settlers.

These squatters' numbers were greatly augmented in the 1970s by indigenous refugees from the guerrilla war raging in Alta Verapaz. Many indigenous communities were attacked by the army on the suspicion that they were aiding the guerrillas. Survivors often took refuge in the mountains. Others were driven from their communities simply because their lands had become potentially valuable for export crop producers who wanted to take control of them.

An example of the latter situation was the notorious Panzos massacre of 1978 in the Polochic valley. This indigenous community had been occupying its lands for more than 80 years and hence its members were entitled to claim legal titles to the community's lands. They petitioned the government for titles and were led to believe that these would be granted by the National Agrarian Transformation Institute (INTA). When they approached the government building in the neighbouring municipal seat expecting to receive titles to their land, however, they were fired upon with automatic weapons by an army detachment. Over 100 were killed. Survivors and their families fled to the mountains.

An indigenous community accessible only by a two-hour hike up a steep footpath high above the Polochic in the Reserve's nucleus area had been established near a small stream by a few dozen families who had fled the Panzos massacre. They had been cultivating subsistence crops there for 15 years and did not dare to return to the valley for fear of the army. An absentee land owner who claimed the land that they had occupied agreed to sell it to them if they made a hefty down-payment in kind, but after they made their payment the owner raised the price far beyond their means. Now the area is to be strictly protected and should not be used for agriculture. The legal owner will presumably be compensated, but not the indigenous settlers.

These refugees had used slash-and-burn farming practices to survive. They had clearly contributed to deforestation in the immediate area, but they had carefully attempted to develop environmentally benign farming and forestry systems within their limited possibilities. FDN promoters sympathetic to the community were attempting to find them a suitable location to move to in the multiple-use area that they could eventually purchase. It was by no means certain that they would be successful.

Looking up towards the Sierra at night at the end of the dry season, it appeared that the whole mountainside was ablaze from fires set by settlers to burn brush and weeds before planting their *milpas*. It is easy to understand why many observers have concluded that peasant squatters were the major cause of deforestation in the area. As the foregoing accounts show, however, the real processes are much more complex and the main culprits are found elsewhere.

There were some 80 small indigenous settlements such as this one on the northern slopes of the Reserve. None had schools or other state facilities as they were illegal squatters and had been considered as guerrilla sympathizers. A well known Guatemalan anthropologist, Myrna Mack, had been assassinated by the army for documenting the harsh treatment many such indigenous refugees endured.

The Mario Dary Biotopo is a bird sanctuary of over 1000 hectares at the top of the Quetzal pass over the Sierra de la Minas just west of the Reserve. It was named after one of Guatemala's most distinguished biologists, who was instrumental in establishing it. He was appointed rector of San Carlos University in the late 1970s, but was assassinated shortly afterwards by a right-wing death squad, apparently because he advocated democratic pluralism. The sanctuary still has impressive tropical cloud-forests and is famous for its quetzals (the national bird). In spite of being an important tourist attraction and foreign exchange earner, the future of this forested reserve may be precarious unless broader socio-political issues in the region are resolved. There were reports of poaching, timber stealing and other encroachments.

On the western boundary of the reserve near the Biotopo, along the Quetzal pass connecting the capital with Alta Verapaz, one finds somewhat different agricultural expansion processes than those described above. Large sheets of plastic protect plantations of vegetables, flowers and ferns from birds, insects and cold in small forest clearings bordering the main road. These produce one of Central America's latest 'nontraditional' agricultural exports. Since the mid 1980s areas of these new agro-exports have been expanding rapidly, especially in the highlands of Guatemala and Costa Rica.

A few kilometres south-east of the Mario Dary Biotopo is the community of Chilasco that borders the western nucleus of the Sierra de las Minas Biosphere Reserve. This village was founded in 1906 when 14 heads of families were awarded land by the dictator of the day in recompense for loyal army service. It now includes 400 families cultivating about 820 hectares. The community is a newly established centre of 'nontraditional' export crop production. There were fields of broccoli, string beans, tomatoes and the like, destined in part for sale and consumption in Guatemala City and in part for export to the United States. This production is organized and financed by a few large buyers based in Guatemala City who are closely linked to transnational importers and exporters. The buyers supply inputs such as improved seeds, chemical fertilizers, herbicides, pesticides, plastic sheets and containers. They also provide technical direction and set quality standards. These costs are later deducted from the value of the crops they purchase from the peasants, whose contracts prohibit them from seeking other buyers. Even so, peasants cultivating broccoli, for example, estimated that with three crops per year they could obtain an income for their own and their family's labour, land and capital of up to 6000 quetzals (about US$1000) annually from one manzana (0.7 hectares). This assumes that there were no crop failures and that market prices hold up, which is not always the case. In fact, markets are volatile for export crops such as string beans, broccoli and ornamental plants.

Incomes in the village were obviously better than those of most residents in settlements near the Polochic valley. Several houses in this community had new zinc roofs or new tiles and thatch. There was a well-kept soccer field, a whitewashed school, newly installed pipes to bring clean drinking water and other signs of relative prosperity. The inhabitants were both *ladino* and indigenous peoples. Some still spoke in indigenous dialect among themselves. Nonetheless, they all wore western-type jeans and shirts that had been made available at low prices by importers of second-hand clothing from the US.

In many indigenous regions, the army, using non-indigenous troops from the east, entered traditional indigenous villages. After abusing or assassinating a few villagers to establish their authority, they would force the men to participate in 'auto-defence' militias directed by the army. In fact these still operated in 1993 near Teleman in the Polochic valley. The army also tried to force villagers to abandon indigenous dress and customs. In areas of guerrilla activity, residents of small villages were required to move to larger settlements under army control. In this particular settlement, however, the acculturation seems to have been more a result of the relative prosperity accompanying nontraditional exports, than of direct coercion.

The new cash crops also have brought problems. Aside from uncertain prices, the peasants have to assume the risks associated with climate, insect plagues and disease. The excessive use of pesticides has led to considerable soil and water pollution as well as the direct personal contamination of workers through exposure. Elimination of many natural predators has left the insects that attack maize free to multiply. Maize production for local use has suffered severely from insect damage as well as from competing demands for land and labour. The more intensive use of land for the new exports has apparently been accompanied by decreasing rates of forest clearance for peasant crops and pasture, at least in this community.

This seems to have been the case too for several communities along the Quetzal pass road that have adopted nontraditional exports. The cropland areas were often expanded at the expense of clearing forests. Nonetheless, the net effect could have been to reduce deforestation rates a little in the region by providing cash incomes and employment and thus reducing pressures to clear more land for maize and pasture.

### *Conclusion*

As in Brazil, state policies (and the absence of policies) were the immediate cause of most undesirable forest clearance. Deforestation resulted from deliberate political choices. Countervailing policies have not been effective on the whole because the state's dominant strategy was one of 'modernization' of a kind most profitable in the short run for its most powerful support groups. These included the traditional oligarchy, the army brass, many foreign and domestic investors, and also the United States' and other foreign 'aid' bureaucracies.

Land tenure institutions played a major role in the deforestation of the Sierra de las Minas. Nearly two-thirds of the Reserve's area was claimed in large holdings by private owners, mostly by non-residents. The remaining 30 to 40 per cent was held by communities or municipalities and by the state. There were some overlapping titles, and property boundaries were seldom marked. Communal land holders suffered the greatest insecurity of land tenure. But land tenure within the Reserve was only a minor part of the land tenure problem. The highly polarized land holding system in surrounding areas was much more important. It denied most indigenous communities secure access to adequate land. It also encouraged large scale capital-intensive agriculture by big owners who employed only a fraction of the potential rural labour force during most of each year, while the majority were landless or nearly landless. It also encouraged the extraction of merchantable timber for the profit of large land owners and timber enterprises but to the detriment of peasant farmers. These often suffered, following logging operations, from water shortages and more difficult access to a wide range of useful products. They were seldom compensated for timber removed even if it came from their own lands. Also, several settlements in the dry Motagua valley had begun to suffer from worsening water shortages. This was one of the factors that helped mobilize public support for the creation of the Reserve.

Indigenous farming systems tended to be more environmentally friendly than those of small *ladino* producers. The latter often aspired to become ranchers instead of farmers. The most damaging farming systems for the forests, however, were those organized by large land holders and investors to produce commercially for export and domestic markets. This 'modern' agriculture displaced some forests directly, but it was most damaging indirectly by excluding peasant producers from access to most good agricultural land.

Population growth had been rapid in both the Motagua and Polochic valleys and on the Reserve's moist northern and western slopes. Immigrants to the valleys had been drawn by new employment opportunities accompanying agricultural modernization, and in the Motagua valley also by growing industrialization and urbanization. Settlers on the southern slopes of the Reserve were few and diminishing in number. They were a minor factor in deforestation compared to logging. Squatters on the northern slopes were mostly indigenous refugees. The areas in which they squatted had usually been logged earlier. They were causing some additional forest degradation by their slash-and-burn *milpa* rotations, but they were also protecting and regenerating forests to satisfy their self-provisioning needs for forest products and for cultural reasons.

## CHINA[11]

About 13 per cent of China's territory was believed to be under forests in the early 1990s (ZGTJNJ, 1992). Its forests are unevenly distributed across the country, depending upon diverse physical features, climates, population densities and economic activities. Most of the forests lie in the eastern monsoon region. In some of the provinces, namely Jilin, Heilongjiang, Zhejiang and Fujian, about one-third of the territory remains under forests. In several provinces of the country's interior, namely Ningxia, Shanghai and Tianjin, forest cover was less than 3 per cent. Tibet, inner Mongolia, Xinjiang and Qinghai have few forests because of their high plateaux, deserts or grasslands (see Table 3.1). Forest types fall into distinct bands from north to south according to climate, with cool-temperate coniferous forest in the north, that are followed successively as one moves south by temperate coniferous and deciduous broad-leaved mixed forest, warm-temperate

11 Unless otherwise stated, material for this section was mostly taken from Rozelle, Lund, Ting and Huang, 1993.

deciduous broad-leaved forest, sub-tropical evergreen broad-leaved forest and tropical monsoon rainforest in the extreme south of the country.

Important tracts of virgin forests still exist in northern China, but most of them are in remote areas with sparse populations. In southern coastal areas, remnants of lush tropical forests are surrounded by economic zones and development districts. Since the

**Table 3.1** *Forest Area and Forest Cover of China's Provinces, 1985*

| *Province* | *Total area* (million hectares)* | *Forest area** (million hectares)* | *Per cent of national total forest**** | *Forest coverage rate (per cent)* |
|---|---|---|---|---|
| Beijing | 1.7 | 0.14 | | 8.1 |
| Tianjin | 1.1 | 0.03 | | 2.6 |
| Hebei | 18 | 1.62 | | 9.0 |
| Shanxi | 15.6 | 0.8 | | 5.2 |
| Inner Mongolia | 128 | 15.23 | 17 | 11.9 |
| Liaoning | 14.6 | 3.68 | | 25.1 |
| Jilin | 18 | 5.8 | 3 | 32.2 |
| Heilongjiang | 46.9 | 15.76 | 8 | 33.6 |
| Shanghai | 0.6 | 8.36 | | 1.3 |
| Jiangsu | 10 | 0.32 | | 3.2 |
| Zhejiang | 10 | 3.37 | | 33.7 |
| Anhui | 13 | 1.69 | | 13.0 |
| Fujian | 12 | 4.4 | 3 | 37.0 |
| Jiangxi | 16.7 | 0.98 | 4 | 32.8 |
| Shandong | 15 | 0.89 | | 5.9 |
| Henan | 16.7 | 1.42 | | 8.5 |
| Hubei | 18 | 3.6 | | 20.3 |
| Hunan | 20 | 6.5 | 4 | 32.5 |
| Guangdong**** | 17.8 | 4.9 | 5 | 27.7 |
| Guangxi | 23 | 5.06 | 5 | 22.0 |
| Sichuan | 57 | 6.8 | 7 | 12.0 |
| Guizhou | 17 | 2.2 | 3 | 13.1 |
| Yunnan | 39 | 9.36 | 10 | 24.0 |
| Tibet | 120 | 6.12 | 4 | 5.1 |
| Shaanxi | 20 | 4.34 | 5 | 21.7 |
| Gansu | 45 | 1.75 | | 3.9 |
| Qinghai | 72 | 0.22 | | 0.3 |
| Ningxia | 6.6 | 0.09 | | 1.4 |
| Xinjiang | 160 | 1.12 | | 0.7 |
| **China** | **955.1** | **116.55** | | **12** |

* Provincial areas taken from Zhou Shunwu, 1992.

** Area estimated by multiplying column 1 by column 4 at provincial levels.

*** Percentages are only included for provinces with more than 3 per cent of national forest area.

**** Includes Hainan.

*Source:* QGLYTJHB, 1989

1970s, many commercially valuable forest resources have come under the control of private entrepreneurs, enterprising state cadres and joint venture operations. Forests in mountainous regions are scattered throughout south and south-eastern China. These areas are generally poor and inaccessible as well as being inhabited by diverse ethnic groups. Many provinces in central China have also maintained or replanted significant areas of forests so that the production of forest products in this region is becoming increasingly important. These geographical variations have to be considered in any analysis of afforestation and deforestation trends in China. They are also important to keep in mind when examining the impacts of public policies affecting agriculture and forests in China's diverse regions.

Afforestation has been very important in China during recent decades. Large tracts of natural forests have been logged, with some of these logged areas converted to other land uses, but the country's total forest area has continued to increase since the 1970s. Nonetheless, as in many developing countries, the protection of newly planted areas is frequently less successful than claimed and several logged forest areas became relatively unproductive.

China has over one billion inhabitants and its economic growth since 1980 is estimated by the World Bank to have been about 10 per cent annually, with per capita GDP reaching US$530 in 1994. As a result of rising incomes and population growth, the demand for forest products has increased dramatically, making the remaining forest resources increasingly valuable. In addition to the growing needs of rural populations, timber is avidly sought for construction and industries. Urban consumers are diversifying their diets to consume many products taken from forest areas.

Under China's centrally planned economy, prices were often set with little consideration of supply and demand before the economic reforms of the late 1970s and 1980s. The timber industry had suffered from low prices and it had not been allowed to retain profits. There were few incentives for sustainable forest management. Also, silviculture and reafforestation had been chronically underfunded. Timber resources were often used wastefully and forest growing stocks had seriously declined. Moreover, several forest areas had been placed under strict protection as national parks and biodiversity reserves, thus reducing the forest areas available for wood production and local uses (Ghimire, 1994). In the early 1980s, reforms in pricing and accounting systems for forest enterprises removed some of these obstacles to more sustainable high-productivity forest management, but managers and bureaucrats still had few positive incentives to increase long-term forest productivity if this implied difficulties in meeting short-term production goals. The present forest policies in China reflect many of these contradictions and tensions.

Land tenure institutions in China have been subject to a series of dramatic changes since the revolutionary forces took power in 1949. A profound land reform redistributed agricultural land held by landlords and rich farmers to landless and near landless peasants. This was soon followed by collectivization, first in large production cooperatives and later into much larger communes. Land use and production were centrally planned and controlled. In an attempt to introduce greater economic incentives and to improve economic efficiency in the use of resources, the family responsibility system was introduced in the late 1970s and early 1980s. Under this system each peasant family was allocated long-term usufruct rights to a particular area and was free to dispose as it wished of any production in excess of quotas that had to be sold to the state at fixed (low) prices. Peasants could sell above quota production in local markets, where prices fluctuated in response to supply and demand and were usually well above official prices.

Although land ownership is ultimately vested in the national state, the rights and responsibilities of the central government and its various agencies, the provinces, the counties, the municipalities and individual producers have been in continuous flux during the last two decades as well as varying considerably among different regions. Uncertainties associated with changing land tenure institutions together with regional variation in the control of forest resources is believed by many analysts to have diminished incentives for long-term sustainable management of forest resources. These uncertainties may have encouraged over-exploitation of forests for immediate gains as well as for their conversion to agricultural and other uses in many areas. This tendency was reinforced by the growing importance of market forces in resource allocation, as well as by the significant immediate revenues that some local authorities can derive from the sale of forest products.

In China's north-eastern region, forest lands are principally administered in state-controlled forest farms, often associated with forest industries. In the south-eastern region, however, forest lands were mostly controlled by the collectives. After the family responsibility system was adopted, some of these forests were allocated for the use of individual families, while others remained under collective control at township or county levels. In the south-western region a mixture of state forest farms and collective control of forest areas predominated (Rozelle, Albers and Li Guo, 1995).

The data on forest areas, volumes of forest growing stock and production in China are not considered to be very reliable. Even more than estimates of agricultural production, they are subject to large margins of error. Some areas that are classified as forest include tree crops such as fruit orchards and rubber plantations. The area of cropland shown in China's official statistics may also be underestimated by between one-third and one-half according to some analysts (Heilig, 1997). If this turns out to be the case, areas estimated to be in forests and in 'other uses' would have to be revised downwards by a corresponding amount, as the estimates of China's total land area are quite firm and inflexible.

The present study was concerned mainly with the semi-tropical forest zones of Yunnan province. In this province, as elsewhere in China, the state's forest policies have been heavily weighted towards increasing timber production. The management of natural forests to maximize timber production and afforestation have gone hand in hand.

### *A case study in Hekou county, Yunnan province*

To examine more in-depth the role of government policy and local deforestation processes, a case study was carried out in Hekou County. This county is located in the south-west province of Yunnan, separated from Vietnam by the Hongjiang (Red) River. Located on a railway line between Kunming and Saigon (Ho Chi Minh City), economic development in Hekou has proceeded rapidly in recent years. In order to attract investment, the county was recently opened to outsiders.

Yunnan had a population of 38 million in 1990, the majority of which belong to the ethnic minorities of Yao and Miao. The population of Hekou county was about 75,000 people, with an area of 1313 square kilometres. The economy of Hekou is predominantly agricultural. While 69 per cent of the gross value of industrial and agricultural output (GVIAO) in the rural sector is from cropping, only 13 per cent comes from industry. Forestry output makes up 8 per cent of GVIAO. The major food crops are rice and maize. Farmers produce sugarcane and fruit as the largest cash crops. Despite their relatively small contribution, forest products have been the fastest growing component of the agricultural output since 1985.

### *Deforestation trends in Hekou*

Although as much as 40 per cent of Hekou was covered until a few decades ago by tropical forests, today only scattered bits of natural forests remain. Tigers, leopards, monkeys, wild boar, deer and a variety of birds once common in Hekou are now nearly extinct. As elsewhere in China, the deforestation of Hekou has resulted from many conflicting policies. Among these were the agricultural policies that stressed grain self-sufficiency, forest policy aimed at increasing county revenue by expanding tree plantations, and land tenure instability.

Deforestation in Hekou is not new. The many ethnic minorities of the area have long practised shifting cultivation to supplement the harvest from their rice terraces. Land is cleared for planting by burning and upland rice is grown for a few seasons, and then the plot is left fallow for seven or eight years. Village leaders claim to burn only grasslands for their swidden crops, since forests are too valuable to destroy. The small areas of primary forest interspersed throughout the swidden grasslands, however, suggest that the area was once covered by forest. Village elders have no recollection of this. The elders report that forest cover (including private tree plantations of teak and Chinese fir) has increased during their lifetime.

The first recorded major episode of deforestation in Hekou took place in the late 19th century when the Kunming–Hena railway was built by the French. Designed to link southern China with the ports and commerce of Vietnam, the railway runs directly through Hekou. Large tracts of forests were cleared to make way for the new mode of transportation and to supply the timber for railway ties and bridges. The railway brought other changes as well. It opened access to Hekou for migration and its population quickly expanded. Large expanses of forest in the river valleys were cleared for paddy rice.

Vast tracts of Hekou's forests were again cleared in the late 1950s to fuel backyard steel furnaces and to make room for the large state collective farms. Because of the scarcity of flat valley land suitable for paddy rice, the collectives cleared forests to establish rubber plantations. Observers agree that most recent forest loss in Hekou took place during this time. To understand better how Hekou's modernization drive had such adverse consequences on its forest resources, it is necessary to examine the county's institutional, forest and agricultural policies.

*Land tenure in Hekou*

Since the 1949 revolution, all land in China has been legally owned by the state. A variety of actors, however, have control over and responsibility for land use. Legally, there are today four types of control over forest and agricultural land: state land, provincial land, village land and private household land. In Hekou, however, land rights are typically less clearly defined than these categories would suggest. While some military and other state agency farms have clearly established their claims to certain areas, most village and household land is distributed under an informal system.

Two large state collective farms cover nearly half of the land in Hekou county. They grow rice in the few river valleys and rubber, pineapple and banana on the sloping hills. Recently, the state farms have implemented the new 'household responsibility system' of farming in which former workers received contracts to cultivate land allocated for their long-term use. In return, the workers are obliged to sell a certain amount of their crop to the collective at state-fixed prices. Any remaining produce can be sold for private profit. While these changes increased the productivity of the collectives, they also weakened the control of the state farms over its workers. As a result, many workers have begun to expand cultivation into adjacent forests that remained under the collective's control.

Due to the myriad of regulations and strict controls associated with the new household responsibility system, many villages, especially those in traditional ethnic communities, do not use formal contracts for private farms. Instead, both agricultural and swidden lands are divided informally among members of a village. Although the farmers have no legal assurances that they will have access to the land in the future, they report having no doubts about the legitimacy of their future claims. Tenure insecurity consequently does not appear to discourage long-term investments in the land, such as the establishment of private tree plantations. This is explained by the fact that the villages in Hekou are quite old, with many generations of the same families working together. Communal cooperation is thus a well-established tradition.

Communal control is also exercised in the management of village forest land. The use of forest land is strictly monitored by the village leaders: permission must be obtained to cut a tree for timber and a small tax is levied. As with agricultural and swidden land, this traditional form of management is effective in controlling forest use. Both village leaders and farmers value the remaining forest for its productive capacity and cultural significance. Village forests appear to have been well-managed and patrolled.

### *Forest policy in Hekou*

As in much of China, the goals within the County Forest Bureau are often at cross-purposes themselves, as well as conflicting with those of other government agencies. For example, the forest bureau is dedicated to preserving and protecting the few remaining natural forest stands in Hekou. To this end, several different county forest reserves have been established. But the Bureau lacks both enforcement manpower and legal recourse against squatters. As a result, the reserves are being whittled away and offer only minimal protection to the forest. The county reserves are frequently encroached by the state farms in Hekou, since they have little incentive to protect the natural forests. To begin, the state farms operate outside of county law and answer only to the central government's agriculture ministry. This ministry's main responsibility is to fulfil production quotas and to carry out expansion plans. In addition, its workers are drawn from all over

China and many have few ties to the local area. Under the new family responsibility system, worker families can expand crop production as much as their labour permits. They too often encroach on county forest reserves. After filling their production quota, they are free to sell their surplus crops on the private market. The state farm administration currently appears to lack control over its individual farmers. Even if the administration had an incentive to curb encroachment on to adjacent county forest reserves, it is unclear that it would have the manpower to enforce it. Hence, the forest management objectives of the state farms may directly conflict with those of the county government.

Private farmers are also encroaching on to the forest reserves, often because of incentives provided by the Ethnic Minority Bureau, a state agency that is responsible for the livelihoods and welfare of the ethnic peoples. Because Hekou County is a Yao Autonomous Region, this bureau has considerable influence in county policy. One of its projects is the '10,000 mu (1 hectare = 15 mu) Policy' in which 10,000 mu of private commercial tree plantations are to be established to increase farmers' incomes. To encourage farmers to participate, the bureau is building several demonstration projects throughout the county to teach farmers how to raise teak, herbal medicines, and spices, in addition to those of the already familiar Chinese fir, banana and pineapple. Lack of tree seedlings and insecure tenure rights have often hindered the success of this programme. Nonetheless, the bureau is a powerful state actor with goals that may potentially conflict with those of sustainable forest management.

The forest reserves are also subject to illegal cutting. At times the illegal logging is unofficially allowed by the forest management bureau in exchange for bribes. In the township of Nanxi, for example, the County Forest Bureau set aside a forest reserve, prohibiting all logging activity. An 'enterprising' township leader, however, cleared one section of the reserve to plant teak and fruit orchards. The forestry bureau officials apparently allowed this because of political obligations and some 'gifts'. Given the chronic underfunding of forest management, their acceptance was hardly surprising.

Forest industry in Hekou operates somewhat more efficiently than the forest department. Forest industry, which consists primarily of a county timber company, has been largely successful in meeting the local government's timber demand. The main reason for this success is that it has better funding and political support than does forest management. This can be explained by the county timber company's importance as a revenue source for the county government. Forest use in Hekou, however, is not fully efficient. Although limits are set by the national forest ministry on the amount of timber that can be cut, they are often disregarded. Recent price increases have created strong profit incentives for local officials to harvest their timber now, since the future of reforms is uncertain in their minds.

### *Agricultural and silvicultural expansion in Hekou*

Feeding its large population has always been a great challenge to China. Yunnan Province is no exception. For a population of 38 million, the province has just 42 million mu (or 2.8 million hectares) of cultivated land, the majority of which is rainfed.[12] Expanding both grain and tree crops has consequently been a priority for provincial leaders. Pursuit of these goals, however, has had dire consequences for Hekou's forests.

To encourage the expansion of grain crops in Hekou, agricultural policy stressed the importance of the county being self-sufficient in food production. The rationale for this policy is clear. Hekou is predominantly mountainous, and transportation both to and within the county is poor. 'Grain-first' policies are consequently seen as crucial to both food security and to raising the standard of living in Hekou and generating income for farm families. To achieve the increase in output, however, more land has been planted while yields per hectare have remained almost constant. Because of the dearth of level valley land in Hekou, this agricultural expansion has often encroached upon forest areas.

Although the soils beneath tropical forests are typically poor for agriculture and cannot sustain crops for more than a few years, recent promotion of chemical fertilizers and pesticides

---

12 This is the official estimate. Some recent publications have suggested that the area of cultivated land in Yunnan may be double this amount (Heilig, 1997). If this is the case, some of the increased cropland area was probably at the expense of forests.

has allowed farmers to continuously plant upland areas even as the soil has eroded. State farms, which occupy virtually all the prime valley land in Hekou, have always had access to fertilizers and pesticides. The recent availability of these farm chemicals to small farmers, however, has allowed a new and growing group to expand agricultural production into upland forest areas.

Commercial tree plantations and fruit and nut orchards have also played an important role in Hekou's development strategy. Lacking adequate groundcover between rows, tree plantations have reduced the ability of forest land to retain water from rain. Water table levels have fallen and soil erosion has increased. Because plantations typically contain just one or two tree species, they lack the biodiversity necessary to provide habitat to animals and to accomplish efficient nutrient cycling. Although they are clearly not a substitute for many crucial functions of natural forests, tree plantations now occupy many of Hekou's forest areas.

Their ecological shortcomings notwithstanding, commercial tree plantations and fruit orchards have been actively promoted in Hekou through trade and price policies. With the current economic reforms introducing more market incentives into China's centrally-planned economy, and also the recent improvement in relations with Vietnam, trade between Hekou and Vietnam is increasing. Since forest resources in the adjacent areas of Vietnam have been largely depleted, market prices for forest products are climbing in Hekou. Increases in the prices of products such as bananas, teak, rubber and pineapple have surpassed those of the price of grain. Farmers in Hekou have switched more to orchards and commercial tree farms to sustain their income. Both private farmers and state farms have expanded teak and Chinese fir plantations. Their effects on forests, however, differ. Because state farms have a limited amount of land, new plantations were often established by clearing forest land. Ethnic farmers, in contrast, have long planted trees on small swidden plots for their own use. Although today they are interested in the cash that timber sales can generate, ethnic farmers still utilize swidden grasslands for their plantations. Their new plantations thus replace neither forest nor agricultural land.

### *Conclusion*

A major lesson from Hekou is that while deforestation is a national concern, neither government agencies nor local people alone are able to reverse the trend. The overwhelming impression drawn from village leaders and farmers is that while only small bits of natural forest remain, the local population does not perceive itself to be negatively affected by the loss. Wild game and plants are not missed. Timber, fuelwood, and cash income needs are met primarily though private tree plantations. Water sources, such as natural springs, and rainfall do not seem to have diminished much as the forest has receded. Because of the rapid growth of grass and shrubs on the mountains, erosion is not a major problem.

The government has also been ambivalent about forest loss. Forest management authorities at the county level claim their responsibility is to protect, not replant, forests, yet they lack the resources to achieve even this limited goal. The Ethnic Minority Bureau has sought to increase forest cover, but only by encouraging more private tree plantations. The state farms operate outside of county law and are interested mainly in short-term profit.

Consequently, it is clear that while private and state farms will plant tree stands for commercial purposes, the protection and regeneration of natural forest must be the responsibility of the state. As China moves toward a market economy, the preservation of natural forests is not now in any private agent's immediate economic interests. As a classic case of a public good, in the present context forests must be protected by the national government.

Another lesson to be drawn is that the economic and forest policy reforms of the 1980s can not by themselves be expected to bring about the sustainable use of China's natural resources and especially of its forests. Timber and agricultural prices are now free to vary with market demand. Trade with neighbouring Vietnam is expanding. Private farmers are being given greater control over production decisions. Forest protection still has a low priority. Producers and officials are more interested in making immediate profits than protecting forests in China, just as they are in other developing countries.

# Cameroon[13]

Cameroon covers an area of 475,442 square kilometres. About 11 per cent of this is in the Sahelian savannah, 30 per cent lies in the higher altitude moist savannah (generally over 400 metres) and 58 per cent is in the moist tropical forest zone. The rest lies in a variety of other ecosystems. This endows the country with a wide variety of natural resources, climates and biological diversity. These geographical zones also have numerous specific demographic, social and land tenure characteristics. The country has ten provinces. Two provinces, Extreme North and North, lie in the Sahelian zone; three, North-west, West and Adamaoua are in the moist savannah zone; and the remaining East, South, Central, Littoral and South-east are mainly in the moist tropical forest zone.

In the far north, semi-desertic conditions exist with little rain. Several dryland crops are grown, such as cotton, millet, sorghum and groundnuts. Rice is also grown where irrigation is available or rainfall is sufficient for dryland varieties. The moist savannah is covered with short grasses and occasional clumps of trees. Rainfall is higher in this zone, especially in the coastal areas and at higher altitudes. Many types of crops are grown both for self-consumption and the market. Cattle grazing and the rearing of sheep and goats are also common. Over 50 per cent of the Cameroon's population, estimated to be 12.2 million in 1992, is concentrated in these two regions, with about 15 per cent of the total land area (Ministry of Planning and Regional Development et al, undated, p2).

The moist tropical forest zone stretches from the Atlantic coast in the west to Equatorial Guinea, Gabon and the Congo in the south, and to the Central African Republic in the south-east. About 35 per cent of this land area is covered by dense humid forest. There are also important swamp/mangrove areas, especially near the Atlantic coast and along Equatorial Guinea. In the mangrove areas, scattered settlements are found, with fishing being the principal livelihood activity. The moist forest areas have been home to many groups of indigenous populations, such as Baka and Bakola. Hunting and forest gathering are the

13 Unless otherwise noted, material for this section was mostly taken from Mope Simo, 1995.

main sources of subsistence for these people, although crop production is also becoming increasingly popular. It is in this region that most logging has taken place in recent years. Two major urban centres, Douala and Yaoundé, and many smaller towns and major agricultural settlements, are found in this region.

According to the FAO's tropical forest assessment, Cameroon's natural forests in 1990 covered some 20.4 million hectares, 43.7 per cent of its total land area.[14] Tropical rain forests made up 30 per cent of this total and moist deciduous forests 55 per cent, with the 15 per cent remainder found in dry deciduous and montane zones. Deforestation during the 1980s was estimated to have been 122,000 hectares annually, which was about 0.6 per cent of its forest area in 1990. It also had 23,000 hectares of forest plantation (FAO, 1993). There had been some deforestation due to human activities long before the colonial period, but deforestation accelerated in the late 19th and early 20th centuries.

Like many West and Central African countries, Cameroon has been rapidly urbanizing. In 1992, about 42 per cent of its population was urban, and in Littoral, North-west and West Provinces over 70 per cent of the population was urban. Average population density in the country is relatively low. There are over 60 inhabitants per square kilometre in the Far North and South-western Provinces, but in the East, Adamwa and South Provinces, population density is only 10 people per square kilometre. This latter area is where most of the remaining reserves of tropical forests lie.

Industrial activities occur in the coastal zone, especially in Douala and Edea, where there is easy access to sea transport. Although some agro-industries have developed in the rural areas, the bulk of the rural population is mainly engaged in crop and livestock production, as well as forestry and fishing activities. Cocoa, coffee, oil palm, cotton, rubber, tea, bananas and pineapples are among the important cash crops grown. Other than in the East, the rural population is deeply integrated in the market system.

---

14 These data differ somewhat from those in Table 1.1 taken from the FAO's *Production Yearbooks*. The authors warn that the data for Cameroon are considered unreliable. Forests include land where tree crown canopy covers more than 10 per cent of the area. Much of this forest may be badly degraded.

Industrialization has remained the top priority of successive Cameroonian governments since its independence. Agricultural modernization, development of cash crops and agro-business have been undertaken within this context. In particular, government development policies have emphasized the need to:

- support the drive for the development of agriculture by local processing of raw materials and the creation of industries that will promote modernization;
- create home-based industries to replace imports, notably of textiles; and
- promote training and employment in the modern sector and particularly in agricultural transformation.

This industrial policy was initially stimulated by rapidly expanding domestic and regional markets. At the same time, the availability of cheap commercial energy from sources such as oil, coal and hydroelectric dams stimulated this process. The country maintained a balanced budget and kept external borrowing low until the mid-1980s, when the dollar-dominated prices collapsed of its main export commodities such as oil, coffee, cocoa, cotton and timber. This depreciation of the dollar against the CFA Franc exposed major structural weaknesses in the economy and especially of its industrial and agricultural development. Since then the country has been plunged into a chronic recession. Livelihoods of the majority of both urban and rural populations have deteriorated and the recent devaluation of the CFA Franc was accompanied by inflated prices of many necessities.

Cameroon is considered by the World Bank to be a middle-income developing country, with a GNP per capita of US$830 in 1992. According to the 1996 UNDP's Human Development Index, it is ranked 127, out of 174. It had slipped into the World Bank's low-income developing country category in 1994 due to negative growth of GDP. Life expectancy at birth was 56 years, and only about 50 per cent of the population had access to health services and safe drinking water. It is believed to have one of the lowest daily food calorie supplies per capita (ie, 1981 calories) among developing countries (UNDP, 1996, p145). The extent of social deprivation obviously varies greatly between rural and urban areas and among different social groups.

Cameroon's economy depends heavily on primary production and primitive agricultural techniques. A lack of external support services for agricultural intensification, combined with high prices for farm inputs and low prices for farm products, left peasants with little choice other than to seek to expand cultivation into forests and other accessible fertile land. In some cases, population pressure played a role, especially in the Far North and Western highland plateaux. But as deforestation in recent years has been acute in the tropical forest zone in the lightly populated south and east, population density is clearly only one factor. Logging for export, the use of agricultural land for cash crop production by para-state companies and wealthier farmers and the urban demand for forest products such as timber, fuelwood and charcoal are amongst the principal proximate causes of deforestation. State forest protection measures, changing land laws and a breakdown in the customary social structure and support systems have indirectly made the lives of many rural dwellers more vulnerable, thereby forcing many to clear the forest. Deforestation processes vary from place to place, as do their social and environmental consequences. These are conditioned both by local factors and the broader social and ecological context. The four case studies, carried out mainly in the moist forest areas, help to bring out several issues and dilemmas.

### *The Cameroon Development Corporation's (CDC) agro-industrial plantations*

Among the different agricultural expansion processes that are taking place affecting forest areas at the local level, the role of agro-industrial plantations is particularly significant. These are large-scale and capital intensive plantations specializing in cash crops for export. The core area of these plantations constituted some 98,000 hectares in 1994. They are chiefly located in four of the country's ten administrative provinces: North-west (Donga-Mantung division), West (Menoua division), South-west (Fako, Meme and Ndian divisions) and Littoral (Moungo division). The largest areas of plantations are found in the South-west and Littoral Provinces, where the soil is rich, rainfall is high and access to transportation is easy. This is also the region where urbanization is rapid and where unique montane and coastal ecosystems exist.

A case study of the CDC in the South-west province showed many interesting features and processes of forest clearance. CDC is the largest, most diversified and one of the most efficient corporations in the country. It is involved in industrial plantations of mainly banana, rubber, oil palm and tea, covering an area of 40,000 hectares. Some of these crops have periodically been highly profitable, but others were much less so. They have differing impacts on the forest area.

There exist three types of farming systems within areas operated by the CDC. First, there are areas directly managed by the company. These are highly mechanized plantations, based on sophisticated methods of land development and crop production. Second, there are areas that were originally developed by the company but considered later on to be uneconomic, frequently due to decline in the price of the crop planted. These areas are leased on a contractual basis to individuals who can muster capital and hired labour. Lastly, the company has sought to encourage smallholders in the vicinity to plant cash crops by providing production inputs in return for their agreement to sell their produce to the company. In addition to these farming systems, it should also be noted that many of the company's workers, totalling some 15,000 in 1994, have tended to maintain small farms in the area by clearing forests in order to supplement their low wages by self-provisioning and sales in local markets.

These different farming sub-systems interact with important consequences for both deforestation and livelihoods. Rubber and oil palm production, for example, have been quite promising in recent years. The company has sought to increase land area under these crops, while keeping the existing area under other crops such as banana. Some 4000 hectares of the recently deforested land in the Boa plain has been planned for industrial plantation and 1000 hectares in smallholdings of plantation crops. The need on the part of the CDC to remain economically viable, on the one hand, and the need of small farmers and CDC workers to increase production for cash, on the other hand, will continue to increase pressures for agricultural expansion in the area.

### *The Southern Bakundu Forest Reserve area*

This case study showed the conflicting priorities between the government and conservation officials, agro-industrial interests, migrant farmers and indigenous population groups. The area was officially proclaimed a Forest Reserve as early as 1940. In order to control 'illegal' cultivation and poaching inside the reserve, more stringent conservation rules were imposed in the 1980s. This was, in part, a result of the availability of financial assistance from foreign conservation agencies.

When the reserve was established there had been little consultation with the local Bakundu people, who naturally perceived its resources as still belonging to them. There have been important in-migrations of peasants and other rural dwellers from elsewhere in Cameroon and from neighbouring Nigeria. Some of them had settled in the area a century ago, while others were relative newcomers. In any event, a large number of them were already there when the reserve was established by colonial authorities. In addition, there are now agro-industrial units including rubber and oil palm plantations and a match-making industry. These companies have received concessions from the government to operate within the reserve. A great many of the workers have become squatters on the nearby land that they cultivate to supplement their meagre wages. In some cases, even civil servants, who are generally badly paid, have joined indigenous peoples, peasants, migrants and workers to clear land for cash crop production.

In 1994, there were some 21 villages within the reserve. Some 27 per cent of the population consisted of indigenous peoples, and the remainder were households that had come from outside the reserve. These migrant farmers play a dominant role in the local economies. The majority of the population, both of local and outside origin, were engaged in commodity production. They produced cocoa and robusta coffee for the market and a variety of food crops such as maize, plantains, cassava, cocoyams, vegetables and fruits. These latter crops were for self-provisioning as well as for sale. Swidden cultivation was practised by all these population groups. This has been quite viable given a relatively low ratio of population to available land. Freehold tenure is becoming increasingly dominant, although the indigenous groups still retain many customary communal land tenure practices such as those of land allocation, cultivation, gender relations and inheritance. Volcanic soils in this region are highly fertile and rainfall is abundant. Thanks to better roads and easy access to ports and major national urban centres, agricultural produce can be easily sold. At the same time, improved road infrastructure has allowed logging enterprises to flourish – often with the involvement of powerful local entrepreneurs and higher national officials.

Good soil and rainfall combined with efficient infrastructure also help explain why agro-industries specializing in rubber and oil palm plantations have been established in the region. As indicated earlier, these industries were normally assigned concessions in areas designated for plantation activities. A decline in world prices for primary commodities such as cocoa and robusta coffee in the 1980s encouraged agro-industries to expand their plantations into new forest areas where soils are initially fertile. This has been the case too with smallholders and richer individual farmers. Tree crops such as coffee are left intact following price falls, while both cash and food crops are planted in newly cleared forest areas. A new variety of cocoa that only takes two to three years to mature has become increasingly popular among well-to-do farmers, who hire wage labourers to clear the forest land. Unemployed youth in the area became a main source of this hired labour force. A mixture of local and external processes have contributed to agricultural expansion and forest clearance in this area. It would be quite meaningless to try to estimate what proportion is due to export crop expansion, to clearance for self-provisioning and local consumption, to commercial logging, to government policies in granting concessions, to road construction and to various other factors, as they are all closely interrelated.

### *The Mbalmayo Forest Reserve area*

The study in the Mbalmayo Forest Reserve area brings out similar socioeconomic and ecological processes to those mentioned above, but there were also additional specific ones. The area is in close proximity to the large town of Yaoundé and there are a number of medium-size towns in the region, but in-migration to the reserve area is relatively insignificant. On the whole, rural population density is very low. There is no sign, for example, of reduced fallow periods resulting from land scarcity (Holland et al, 1992). For much of the rural population, a shortage of labour rather than land is a major problem. Deforestation is perceived as an opportunity to improve livelihoods rather than as a socio-environmental problem. For peasant farmers, there are no direct costs but only benefits from forest clearance. More labour is also needed because land quality is generally poor in the region, and much of the best fertile land is now included inside the reserve where peasants' access is restricted.

Traditional land tenure arrangements still prevail among the original inhabitants although, encouraged by state policies, these are rapidly being replaced by freehold land rights held by individuals or corporations. A typical farm, operated by an indigenous or migrant farmer, comprises of a small perennial home-garden, a perennial tree-crop area of cocoa under the forest canopy interspersed with fruit trees and bananas, and a food crop area involving the rotation of cropland with second growth forest cover. Food production activities take place mainly within the reserve because of the fertile soil. In any event, much of this land was already used by local inhabitants before the reserve was established. Although some new areas are cleared in an attempt to grow plantain and other food crops for the market, on the whole, peasants clear fallows in parcels that they had farmed before the reserve was established. This is because land clearance is easier in secondary forest areas than in dense virgin forests. Local level pressure by peasant farmers on large-scale deforestation in the area seemed to be relatively low.

### *The Kilum massif area*

The case study of the Kilum montane agro-ecological zone looks at changing patterns of farming systems, local social structures and related processes. It brings out processes that contrast with those seen in the above case studies in low-land tropical forest areas. Social and ecological characteristics in this area, however, have much in common with highland areas in East Africa and elsewhere (Barraclough and Ghimire, 1995). Soil erosion and the deterioration of watersheds are visible in some locations. Bush fires, shortening of fallows and limited extension of cultivation into forest and brush areas can also be observed. Fertile land is scarce and highly sought after. There are a few areas where population density has reached over 100 persons per square kilometre, although the average density in the massif area as a whole is under 50 persons per square kilometre.

Traditionally, land use systems tended to be highly sustainable. Most households practised integrated agro-forestry practices, and fallows were long. Livestock raising, which is an important economic activity in the region, was based on transhumance. The traditional land tenure system guaranteed access to land to all the community members and social inequalities were less pronounced.

The arrival of cash crops such as coffee and the commercialization of livestock brought about important social differentiation. The richer and more successful farmers were able to hold more productive plots, consolidate holdings and take better advantage of infrastructure and state support services. Apparently, state extension services coordinated by the North West Development Authority (MIDENO) on the whole were highly supportive of average peasants' livelihoods. Some call this programme 'one of the most successful integrated rural development projects in the country' (Mope Simo, 1994, quoted in 1995, p141). It provided credit, improved varieties of seeds, technical assistance and better marketing facilities. But recent structural adjustment programmes have led to the suspension of the activities of MIDENO, due to lack of funds. At the same time, the government and international conservation agencies have sought to include much of the remaining forest and other mountain common property resources into a comprehensive protective regime, thus curtailing peasants' customary access to these areas. Withdrawal of state support for agricultural intensification and the growing control by conservation agencies over the remaining uncultivated land areas left peasants in a very difficult situation.

### *Conclusion*

It can be seen from the above discussion that local level processes affecting agricultural and forest land use in Cameroon vary widely. In general, it appears that population growth and poverty are inadequate explanations of increased deforestation. Population growth has been very rapid in urban areas, but within the rural areas studied, recent in-migration has been low, as has the rate of natural increase. During the last decade, Cameroon's agricultural population increased by only 13 per cent while its total population grew by nearly one-third (FAO, 1995).

The greatest confusion lies with land tenure arrangements. The government's 1974 tenure reform favouring a freeholding system of individual and corporate private holdings has caused disruption of traditional communal land tenure practices, but it has not been able to replace them. The state's control over customary common property resources such as the forest areas is frequently contested by local communities. This generates little local interest in protecting forests and soils, while the state agencies do not have the capacity to do this by themselves.

The government has many contradictory agricultural and forestry policies. On the one hand, it is actively promoting agricultural expansion for cash crop production and export. On the other hand, it has sought to protect forest and water resources by creating parks and reserves and restricting access to them by customary peasant uses. At the same time, production within the peasants sector is increasingly directed towards markets over which producers have no control. The government's development strategies have actively promoted this process. The dilemma with respect to the reliance on the market is that both the rise and fall of commodity prices have tended to put greater pressures on cultivatable forest areas. At local levels, agricultural intensification, improved social services, clear and equitable land tenure rights and greater local participation in the protection and use of natural resources are crucial. Such policies could help in reducing undesirable agricultural expansion into forest areas that for various reasons should remain forested.

## MALAYSIA[15]

Malaysia is composed of three major regions: Peninsular Malaysia, Sabah and Sarawak. It has a total land area of 33.2 million hectares, of which the peninsula covers about 40 per cent, Sabah 22

15 Unless otherwise noted, the bulk of the information used in this section is derived from Jomo K. Sundaram and Chang Yii Tan, 1994 (draft).

per cent and Sarawak 38 per cent. In the early 1980s, about two-thirds of the national territory was under natural forests.[16] Another 3 million hectares were under rubber and oil palm, thus covered generally by perennial vegetation. Regionally, Sarawak had nearly half of the nation's forest areas; the peninsula held about one-third; and the remainder were in Sabah. These three regions differ enormously in their land use history, deforestation patterns, demographic characteristics and socioeconomic conditions, although they share many similar characteristics as well.

Malaysia's population in 1997 was about 21 million. Approximately 80 per cent of the population was concentrated in the peninsula. The national population growth rate was 2.4 per cent per year. Sabah's higher rate of population increase at 3.7 per cent per annum includes considerable net immigration from neighbouring areas in the Philippines and Indonesia. Internal rural to urban population movements have been rapid since the 1960s. In 1994, some 53 per cent of the country's population was urban. This urbanization was stimulated by rural poverty and a high demand for unskilled labour in urban areas. In 1990, some 29 per cent of the rural population, compared to 7 per cent in urban areas, were believed to live in poverty (UNDP, 1996). Swidden farming still persists in parts of Sabah and Sarawak and is often blamed for deforestation. As will be seen later, it is a factor in some places, but a relatively insignificant cause of forest clearance compared to many other socioeconomic processes.

In recent years, the percentage of the labour force in the primary sector consisting of agriculture, forestry mining and fishing has been in steady decline. In 1992, it was about 27 per cent of the population, as compared to 63 per cent in 1960. Industrial and service sectors, on the other hand, have expanded remarkably. In 1960, these activities employed 12 and 25 per cent of the labour force respectively, while in 1992, industries employed 23 per cent and services employed 50 per cent (UNDP, 1996, p168).

Malaysia's per capita income was US$3140 in 1993, which puts it in the World Bank's upper-middle-income group of developing countries. The distribution of income, however, was among the most concentrated of South-east Asian countries. Between

16 The FAO's tropical forest assessment estimated only 53.5 per cent for 1990 (FAO, 1993).

1981–93, for example, the lowest 40 per cent of the households in the country were estimated to have 12.9 per cent of the national income, as compared to 15.5 per cent in Thailand, 16.6 per cent in the Philippines and 20.8 per cent in Indonesia (UNDP, 1996, p170).

Rapid industrialization took place in the 1970s and 1980s. The industrial sector is currently receiving a high priority within the government's development strategy. Primary commodities, however, are still the main export earners. Petroleum, timber, oil palm and rubber constituted the principal commodity exports. Peninsular Malaysia is by far the most developed in terms of infrastructure and manufacturing. For Sabah and Sarawak, petroleum and timber have been the backbone of the economy, with timber exports to Japan being particularly important. There is little doubt that the Malaysian economic growth in recent decades has been heavily dependent on the exploitation of natural resources. Between 1971–89, for example, resource rents from timber and minerals were estimated to amount to one-third of Malaysia's gross investment.

The contribution of natural resource exports to economic growth is likely to be substantial for several years to come, but the government's highest priority is clearly the development of the manufacturing sector. Besides refined petroleum products, the government plans greater processing of natural resources such as the production of veneer and plywood instead of the export of raw timber (ie sawn logs). Rubber and oil palm plantations are becoming more mechanized and economically efficient, although the social conditions of many plantation workers continue to attract much criticism. The establishment of new agro-export plantations by clearing forests has declined recently due to several reasons. Among these are the scarcity of easily accessible and productive land, rising costs of land development and growing criticism of deforestation activities by environmental NGOs and others in the community.

There have also been several official attempts to protect forests through legislation, such as the establishment of strictly protected parks and nature reserves as well as the introduction of scientific forest management in reserve forests. However, in many cases, political economy realities have thwarted these initiatives.

The autonomous political power of each Malaysian state in the control of its natural resources limits the scope of the federal government's authority over forest use. So too do coalitions between powerful business and political élites protecting their special interests. These frequently conflict with the interests of local low income producers attempting to protect and improve their livelihoods. These conflicts have rendered many apparently enlightened public policies ineffective, if not obsolete. There are numerous contradictory processes at state and local levels. The fate of the remaining forests is far from certain, even though direct pressures on the forests, arising from agricultural expansion involving both plantation and peasant sectors, have receded in recent years.

### *Complexities of land use change and deforestation at the state and local levels*

In Peninsular Malaysia, the major processes of deforestation can be traced from the beginning of the 20th century. They were directly related to the expansion of export-oriented plantation agriculture. There was a rapid increase in highly profitable rubber plantations until the Second World War. These plantations were operated principally by foreign companies. After independence, in the 1960s and 1970s, oil palm plantations expanded. About 1.12 million hectares of forest area was cleared for oil palm, and 226,700 hectares for rubber between 1966 and 1984 alone. Mixed horticulture, other crops and paddy also expanded in terms of the area brought under cultivation. Shifting cultivation comprised a very small percentage of overall land use in Peninsula Malaysia.

After 1956, the Federal Land Development Authority (FELDA) played an important role in clearing forest areas for agricultural use. One of its primary missions was settling small landless farmers. It also attempted to consolidate a Malay peasantry that would support the government against the communist-led insurgency in the 1950s. By the late 1980s, FELDA had developed nearly one million hectares of agricultural land, principally for the cultivation of oil palm and rubber. There was a steady increase in agricultural area after the 1950s, reaching nearly 4 million hectares in 1986. The area devoted to the production of rubber subsequently stabilized, but the area under oil palm continued to increase until the late 1980s.

Agricultural expansion and deforestation patterns in Sabah appear somewhat similar to those of the peninsula, but on a much smaller scale. Commercial logging, however, was a very important cause of forest clearance and degradation in Sabah. The British Borneo Timber Company had a monopoly of Sabah's timber utilization between 1919 and 1952. After the Second World War, three large foreign companies and eight local firms expanded logging operations. The Sabah Foundation was established in 1966 to exploit timber resources, as well as to develop the forestry sector along silvicultural lines. The state government granted the Foundation 855,000 hectares of forest area in 1970. Since the 1980s, several large investments by commercial companies have been made in the forestry sector, including forest plantations.

About 9 per cent of Sabah's land area was brought under cultivation by 1989, with most of this area being planted with oil palm, rubber and cocoa. Shifting cultivation accounted for 2.7 per cent of the agricultural area. The land area utilized for the production of annual food crops such as paddy expanded very slowly between 1960 and 1990.

Agriculture has continued to be an important component of Sabah's economy. Large-scale plantations, mainly of oil palm and cocoa, expanded rapidly in the 1980s. Sabah's state development agencies such as the Rural Development Corporation (KPD), which also promoted cash crop production among smallholders, promoted some deforestation, but the areas were insignificant compared to those on the peninsula. For example, KPD cleared only 33,000 hectares of forest area between 1977 and 1988, principally for planting cocoa, coffee and some shorter-term crops.

In Sarawak, logging is considered to be the major direct cause of deforestation. The World Bank suggests that of the 3.1 million hectares of the country's forests logged in the 1980s, 2.3 million hectares were in Sarawak (World Bank, 1991, p4). A Malaysian government source indicated that by 1990, about half of Sarawak's 8.3 million hectares of forest had already been logged to remove the most valuable timber. Many specialists believe that such logging operations damaged from 30 to 70 per cent of the remaining trees. The World Bank estimated that by the end of the 1980s the logging industry in the state employed about 60,000 people which was about 40 per cent of the workers employed by the timber industry in the whole country (World Bank, 1991, p7).

The adverse impact of logging as carried out in Sarawak on the environment and livelihoods of local indigenous peoples was widely publicized by many non-governmental environmental and development organizations. Since 1987, negatively affected tribal peoples have staged blockades of roads used to extract timber, and other protests against logging companies have become widespread. The government has been forced to reduce logging activities in Sarawak somewhat, taking into account recommendations made by the International Tropical Timber Organization (ITTO).

Shifting cultivation is widespread in Sarawak. By the late 1980s, it was estimated to occur on about 18 per cent of the state's total land area. It was, however, mostly concentrated in dryland hill forests (which comprised 56 per cent of the state's forests). Clearance of tropical forests for swidden cultivation covered less than 4 per cent of the state's total land area. About 20 per cent of the state's population of only 230,000 people were estimated to have been engaged in this agricultural practice. Swidden cultivation had existed in Sarawak for many centuries. As noted in earlier case studies, swidden cultivators usually prefer to clear secondary growth forests for their rotation as these require much less labour to prepare than do virgin forest areas.

Other agricultural land uses are minor, accounting for less than 4 per cent of total land use. Estate and plantation areas accounted for about 1 per cent and agricultural smallholders for another 2.7 per cent of the state's land area. By 1988, the principal five crops (ie, rubber, oil palm, cocoa, paddy and pepper) covered 456,000 hectares or under 4 per cent of the state's total land area. The bulk of agriculture in Sarawak has remained smallholding, rather than large plantation, agriculture. Plantation agriculture is a relatively new phenomenon to Sarawak, although in recent years the state government has provided certain economic incentives to develop this sector on a more significant scale.

### *Conclusion: recent trends and policy issues*

Malaysia's experience with deforestation and agricultural expansion has been somewhat different from that in many developing countries. Rural poverty, skewed landownership and population growth have apparently not been major factors in driving deforestation processes in Malaysia. Moreover, the pace of agricultural expansion has recently slowed. New non-agricultural pressures on forest areas have emerged from logging (mostly for export), the development of infrastructure, dams, tourist resorts, land reclamation and aquaculture projects. The government's recent drive towards greater urbanization and industrialization, however, has resulted in many agricultural lands being converted to urban uses such as housing, commercial real estate and physical infrastructure.

As far as the agricultural sector itself is concerned, further expansion is no longer forecast to lead to major deforestation. Smallholder agriculture, which was actively promoted after independence through many rural development programmes, has expanded only marginally in recent years. There is now some preoccupation that agricultural production may be constrained by a lack of labour due to growing rural-to-urban migration, and that some good cropland may be left uncultivated. The pace of shifting cultivation too is unlikely to grow. Most of those engaged in this farming system seem to have enough land for maintaining sustainable fallow periods. This is obviously likely to be interrupted if timber operations and cash-crop plantations continue to encroach on shifting cultivators' territories and if the government establishes new strictly protected forest reserves or national parks at shifting cultivators' expense. The new generation of swidden cultivators, however, tends to seek wage employment in settled agricultural areas and urban centres where wages are more attractive. Regarding plantation agriculture, the government appears to be slowing down investments in public sector land development projects, but encouraging the private sector to participate more in crop production as well as agro-processing. This sector, on the whole, has remained economically competitive. Plantation agriculture remains significant and may continue to expand in Sabah and possibly in Sarawak as well. But there have been few recent important expansions of commercial agriculture into Peninsular Malaysia's forest areas.

As mentioned above, pressures from abroad and national NGOs have had some positive impact in slowing careless logging. Nevertheless, logging still provides an important source of revenue for the Sabah and Sarawak state governments. They are thus unlikely to stop granting forest concessions. This is especially the case in Sarawak, where logging is not only an important source of state revenue but also of political patronage. The state governments, therefore, have considerable autonomy in deciding how forests should be used. At times, there have been powerful political pressures from within the states for legal and illegal logging. Their fiscal structure, however, leaves them highly dependent on the federal government for much of their other revenues. This could enable the federal government to use its financial contributions to dangle budgets as a carrot to induce state governments to implement improved forest protection measures.

Malaysia is now attempting to become a 'developed nation' by the year 2020. To realize this, it is estimated that the country will require an economic growth rate of at least 7 per cent per annum. It is doubtful that agriculture will be able to contribute to rapid economic growth at rates approaching its contribution in the past. Indeed, much of the country's productive resources and policy priorities have already been shifted to the manufacturing and service sector. The relative importance of agriculture in the national economy will probably continue to regress. Since there are few lowland forest areas left to exploit where agriculture is promising, there will be negligible pressure on these forests for crop production purposes. As for the upland and relatively inaccessible forest areas, it is highly improbable that agricultural expansion will continue to be as important as in the past. Deforestation is likely to increase in the short run, however, because of commercial logging and pressures from other non-agricultural uses such as infrastructure and urbanization.

Current deforestation processes may be controlled if there is greater commitment by citizens and authorities alike to environmentally and socially sustainable development. This implies, among other things, the protection of forest-based customary livelihoods. In recent years there has been increasing local level awareness of these issues. Several NGOs have mobilized support for the protection of Malaysia's forests and its forest dwellers. The government has not been able to neglect totally local as well as international pressures to stop undesirable deforestation in the country. Many formal policy measures that have been developed, however, remain to be effectively implemented. Pressing environmental issues frequently receive scant attention at federal and state levels. Political commitments to deal with them constructively will require the mobilization of popularly based social forces.

Before attempting to make generalizations based on these cases, it is instructive to revisit them, focusing on the role of international trade in promoting or deterring agricultural expansion and deforestation. This is done in Chapter 4.

# 4 Linkages with International Trade

## Introduction

Concern has been expressed in various international fora and publications that the trade and agricultural policies of industrialized countries of the North were directly contributing to undesirable deforestation in the South. This hypothesis suggested that reforms in the international trading system, and of trade and agricultural policies in the industrialized countries of the North, could contribute to slowing unnecessary tropical deforestation. The authors of this book attempt to examine this issue in the light of the case studies that were summarized in Chapter 3.

The methodological difficulties of analysing linkages between trade and the agricultural policies of developed countries, and deforestation processes in developing ones, are formidable to say the least. Nation states and their economies can usefully be envisioned as complex, open, interacting systems (García, 1984). Trade among nations implies exchange of goods and services across national frontiers, but this is only one of the many ways in which complex societies interact. Deforestation processes stimulated by foreign trade in a particular tropical country are in part shaped by that country's socioeconomic and political structure at a given time as well as by many other factors, such as its bargaining power vis-à-vis its various trading partners. Linkages that appear clear and firm in one period may be very different and perhaps contradictory in another. Simplistic models showing correlations between tropical deforestation and foreign trade can be dangerously misleading as guides to policy in both developing and developed countries. They are unable to take the complex interrelationships between dynamic open sub-systems into account.

Often it is impossible to estimate even roughly the extent to which agricultural expansion into forest areas is driven by demand for agricultural exports in contrast to demands for agricultural products for domestic consumption and industries. These difficulties increase with a country's size and its level of development. China and Brazil, for example, are large countries with important and dynamic industrial sectors. The export of agricultural output from the Amazon region to Brazil's industrialized south, and from China's south-western Yunnan province to its more industrialized east as well as to urban areas of Yunnan itself, far overshadows the importance of farm exports abroad. Moreover, the import content of both exports and imports becomes increasingly difficult to identify as national economies become more integrated and technologically sophisticated. Malaysia's exports of oil palm, rubber, timber, minerals, petroleum and other primary commodities in the mid-1990s were nearly equalled by its exports of manufactured products. Guatemala's and Cameroon's agricultural commodity exports have a high import content in addition to a significant share of many export crops being destined for domestic uses.

In some cases, expanding commodity exports seem to have been a principal factor driving recent deforestation. Forests may be cleared to make way for export crops and may be cleared or badly degraded by the export of timber. These activities bring migrations of workers and settlers into forest areas who, in turn, undertake forest clearance for food production and other purposes. More indirectly, export crop expansion and modernization in non-forest areas often accelerates the displacement of peasant producers and rural workers made redundant by greater mechanization. Some of these displaced peasants and workers migrate to forest frontiers. In the case study countries there were few reliable data that would permit even very rough approximations of the quantitative importance for deforestation of diverse interacting and overlapping deforestation processes. Moreover, commodity exports may be accompanied by increasing food imports that may, in turn, temporarily at least, diminish pressures to clear forest for food crops.

The dynamics and impacts of forest clearance are to a great extent determined by a complex series of public policies and social institutions that have evolved in each country as a result of historical processes that are always to some extent unique. Cross-country correlations between deforestation and abstract national level variables such as income levels, trade policies, foreign indebtedness, food crop self-sufficiency, human rights, demographic trends and the like are always open to very divergent interpretations. The FAO's publication of data estimating changes in forest cover during the 1980s in 89 tropical countries stimulated a number of such cross-country statistical analyses (for example, Brown and Pearce, 1994). Not surprisingly, relationships between deforestation and other macro-variables turned out to be weak, volatile and often contradictory. Considering the data and the methodologies used, it is hard to imagine how they could have been otherwise.

In the real world, trade is mediated by social institutions in which some groups gain from exports and others are likely to become worse off than before. Trade is seldom the win-win proposition depicted in elementary economics texts. Much depends on the time period considered and the criteria used in judging how it affects different parties. In looking at linkages between deforestation and international trade in the case study countries, their impacts on the livelihoods of the low income groups affected during the short and medium terms were among the principal evaluation criteria employed in this book.

## BRAZIL

As was seen earlier, the expansion of sugarcane plantations and later of coffee and cacao primarily for export was a principal factor contributing to Brazil's economic and demographic growth from early colonial times until the mid-20th century. Commodity exports were estimated to have accounted for from one-tenth to one-third of the country's economic growth during the 19th and early 20th centuries (Bulmer-Thomas, 1994). This agricultural expansion stimulated by export markets was also responsible for the clearance of the major portion of Brazil's once extensive Atlantic coastal forests. Much of this previously forested area, however, was suitable for sustainable agricultural uses, with deep soils, gentle topography and adequate rainfall.

One cannot judge the virtual destruction of most of Brazil's Atlantic coastal forests to have been more detrimental to the possibilities of eventually achieving sustainable development than was forest clearance in much of Western Europe or in the eastern

United States. As in many now industrialized countries, forest clearance may have been unnecessarily wasteful and many areas were cleared that in retrospect might better have been left forested. Nonetheless, export-led agricultural expansion played a crucial role in Brazil's development into what is now an upper-middle-income newly industrialized country. The social impacts of this agricultural export-led development, however, were catastrophic for large sectors of the country's population. Its indigenous peoples were nearly exterminated by disease and harsh treatment, as they were in most of the Americas. Slaves brought from Africa seldom enjoyed improved livelihoods, nor did most of their descendants. Many immigrants from Europe, the Near East and Japan prospered, but some experienced grinding poverty for generations. Poverty in Brazil's north-east, where sugar exports began, has become notoriously severe and persistent. Land degradation in this sub-humid tropical region was accelerated by careless deforestation followed by unsustainable agricultural practices that contributed to worsening poverty and massive out-migrations, especially during drought years.

The main explanations of massive and persistent poverty in Brazil's relatively prosperous national economy, however, have to be sought not from trade and demographic growth, but from institutions and policies that excluded the poor from opportunities to improve their livelihoods. Slavery and its aftermath strengthened a constellation of such institutions. These included a land tenure system that denies most of its participants access to sufficient resources for self-provisioning. The political institutions required to maintain the control of land and labour by a class of large property owners excluded most of the poor from meaningful participation. International trade contributed to this situation, but only indirectly. Colonial élites and their successors deliberately constructed an institutional framework that allowed them to control most of the benefits from trade in partnership with counterparts abroad who also benefited. Persistent and often worsening poverty during periods of booming agricultural exports were accompanied by rapid deforestation. These negative impacts associated with trade could not have been dealt with merely through more socially and environmentally sensitive trade policies in countries receiving Brazil's exports. Such policies might have helped in some instances, just as Great Britain's banning of the slave trade early in the 19th century probably contributed to slavery's eventual demise in the Americas several decades later, but this was only one factor among many others.

Present-day deforestation in Amazonia and in one of Brazil's few remaining Atlantic forests examined in the local level case studies is for the most part only indirectly linked with foreign trade in agricultural and forest commodities. Cattle ranching is directly associated with current recent changes in forest land use as most of it was converted to pasture. In 1994, however, only a little over one-fourth of Brazil's merchandise exports were agricultural products. Only 3 per cent of these exports consisted of meat and live cattle, most of which were being produced in the South rather than in Amazonia (FAO, 1995). State subsidies and related policies drove most deforestation in Amazonia rather than international markets for meat and other agricultural commodities. This was clearly brought out in the five local level cases. Nonetheless, a few wealthy cattle ranchers linked to export markets probably exercised excessive influence in bringing about such socially and environmentally harmful policies.

In the Mato Grosso cases there were some direct foreign investments in agro-export enterprises. These mostly failed eventually, but in any event the foreign investors were apparently more attracted by the prospect of short-term subsidized profits accompanying speculative activities than in longer-term agro-exports. The spectacular boom in soya exports after the 1950s was in part driven by the European Community's common agricultural policy (CAP). While there was some soya production on formerly forested areas in Mato Grosso, this was not a significant factor in its deforestation. The expansion and mechanization of soya production for export in Paraná and other southern states indirectly contributed to deforestation in the Amazon states, however, by making much of their rural population redundant. Many migrated to the Amazon region attracted by promises of land or jobs.

Speculative foreign and domestic investments in Amazonian forest lands were also stimulated by the region's huge timber and mineral reserves, reputedly the largest in the world. The economic returns for Brazil from these investments still lie far in the future, although speculators often gained handsome short-term financial rewards related to state subsidies. Timber and mineral exploitation for export could easily become a much bigger factor than agricultural expansion in forest clearance in the Amazon region in the future unless adequate measures are taken by the state to compel investors to bear the costs of social and ecological externalities resulting from their activities.

## Guatemala

The impacts of foreign trade on agriculture and forests in smaller Guatemala were much more salient than in big Brazil. In the former, 1994 agricultural exports amounted to 7 per cent of its GDP, while in the latter they were only 2 per cent. Moreover, in Guatemala agricultural exports made up 63 per cent of total merchandise exports, in contrast to only 27 per cent in Brazil. Guatemala's officially recorded agricultural exports were primarily coffee, sugar, bananas, fruits and vegetables. Cotton had been an important export in the 1970s but had declined sharply in the 1980s. Forestry exports were officially negligible, although illegal exports of timber through Mexico and Belize were probably significant. Moreover, Guatemala was importing nearly one-fourth of its food supplies. In the early 1990s cereal imports alone amounted to some 40 kilograms per capita (FAO, 1995).

The preceding chapter brought out the leading role of agro-export expansion in stimulating deforestation throughout Guatemala's history and especially after the mid-1950s. Agro-exports have also been the principal motor driving the country's economic growth. As was seen earlier, growth of GDP in the 1960s and 1970s was over 5 per cent annually, falling to less than 1 per cent in the 1980s when many of its agro-export markets virtually collapsed. The earlier discussion also emphasized the serious livelihood crisis faced by Guatemalan peasants, and especially by the indigenous majority. This had been in part provoked by displacements of peasants as a result of agro-export producers taking possession of their traditional lands.

The United States' aid, trade and agricultural policies have been very influential in shaping the Guatemalan state's development strategy. These US policies had been influenced by the perceptions of US commercial and strategic interests held by US exporters, importers and investors. These had particularly negative consequences for most peasants in Guatemala because 'aid' policies that could have contributed to more equitable development had been subordinated to the US government's perception that popularly based struggles for land and other social reforms were manipulated by the USSR and Cuba for their advantage in the 'Cold War'. This led to massive US military and economic 'aid' for Guatemala, but with the primary objective of supporting the government in its conflict with guerrilla forces. Social and environmental concerns were given a very low priority. As was seen earlier, the immediate origin of the civil conflict in Guatemala had been the US-engineered military coup of 1954 that reversed the Arbenz administration's initially successful agrarian reform.

The US agricultural strategy for Guatemala was to promote and modernize agro-exports, to encourage colonization of the still forested agricultural frontier as a substitute for land reform and to make Guatemala a better market for US surpluses of cereals, dairy products and other agricultural exports. International financing was made available through USAID, the Interamerican Development Bank, the World Bank and other sources on attractive terms. The government was able to improve roads and other infrastructure, provide cheap credits for agro-export producers and commence the colonization programme mentioned in the previous chapter. In addition, the US government encouraged the US private sector to invest in Guatemalan agro-exports by providing it with insurance against many risks.

Initially in the 1960s, US policies promoted Guatemalan livestock and cotton exports to meet rapidly growing demands in the US for these commodities. The US government also enlarged Guatemala's sugar quota after eliminating Cuba's, and US buyers purchased a major share of its coffee and banana exports. When terms of trade turned against these commodities in the 1980s, the US promoted 'non-traditional' exports of fresh fruits and vegetables, flowers and ornamental plants. On the whole, the expansion of agro-exports contributed to accelerated deforestation both directly and indirectly, although fruit and vegetable exports in some cases may have slowed deforestation for maize and bean production by providing alternative income sources for a few peasant producers. The impacts of agro-export expansion on livelihoods were negative for most, but not all, peasant producers and rural workers.

As in Brazil, these negative impacts associated with trade have to be explained principally by policies and social institutions that excluded most of Guatemala's poor peasants and workers from the benefits that are presumed to flow from agro-exports and economic growth. As already discussed, its land tenure system is one of the most polarized of any country, leaving the rural poor with almost no resources or opportunities. Unless public policies and social institutions are profoundly reformed, widespread poverty will persist whether agro-exports are booming or contracting. So too will undesirable deforestation.

Guatemala became increasingly dependent on food imports after the 1950s. Food aid and subsidized commercial imports of wheat and many other foods rose rapidly. Food imports were financed under various titles of US Public Law 480. Guatemala had been practically self-sufficient in cereals in the early 1950s, but by 1990 it was importing nearly one-quarter of its cereal consumption. Most of its urban population and many of its rural people were heavily dependent on imported food for their diets.

This food import dependency had contradictory consequences for agricultural expansion and deforestation. Less land was required for food production in Guatemala than would have been the case if it produced most of its own food supplies. The area in maize increased more slowly than did population, while average yields remained low and stagnant. Many peasants who could be producing food remained underemployed as they lacked both land and markets. Food imports contributed to agro-export expansion by making it more feasible and profitable for commercial producers to concentrate on export crops. Cheap food imports also contributed to maintaining Guatemala's archaic land tenure system and other quasi-feudal social institutions by easing political pressures on the government to carry out needed reforms. As is frequently the case, food aid, together with agro-export dependency, have indirectly helped to perpetuate the poverty, repression and environmental degradation they were supposed to reduce. This is not a problem that can be solved through more enlightened trade policies either in the US or Guatemal could help.

## CHINA

China has nearly eight times Brazil's population, inhabiting an area that is only 12 per cent greater, part of it desert. Its total GDP, however, is only about the same as Brazil's, although it has been growing much faster than Brazil's in recent years. Given its size, its population, its low per capita income and its fast economic growth, China can be expected to be an increasingly important actor in international trade. These same factors, however, suggest that foreign trade may have a minor role to play in Chinese agricultural expansion and deforestation.

The available data support this hypothesis. In 1994, China's total agricultural exports of an estimated US$14 billion amounted to about 2.7 per cent of its GDP and to only 6.7 per cent of its total merchandise exports. Its total merchandise exports, however, amounted to 38 per cent of its GDP. Its agricultural imports were slightly larger than agricultural exports. Forest products comprised only half of 1 per cent of its merchandise exports, but over 2 per cent of its merchandise imports. The estimated value of China's imports of forest products was nearly twice that of its cereal imports. China is one of the world's largest traders of agricultural and forest products in spite of its low per capita income, but it is also one of the countries in which the foreign trade of agricultural and forest products is the most insignificant in comparison to trade in domestic markets.

Other than in limited areas, the international trade of agricultural and forest products does not play a direct role in influencing agricultural expansion and deforestation processes. China's large net imports of forest products have helped to diminish pressures to overexploit its remaining natural forests. They have also contributed to increasing demands for timber from exporting countries. China and India are the world's largest tea producers and China is a primary tea exporter, with tea comprising about one-fourth of all its agricultural exports in the early 1990s. Nonetheless, there has been little expansion of areas under tea production in recent years. Net imports of cereals in the 1990s may have contributed to reducing expansion of these crops into forested areas. However, greater internal trade in cereals, stimulating specialization of crops in regions most suitable for their production, has a greater potential for contributing to the preservation of remaining forest areas.

The forced opening of China's markets by western imperial powers and Japan in the 19th and early 20th centuries stimulated demographic changes and agricultural expansion that often occurred at the expense of China's forests. Current rapid growth of manufactured exports and foreign investments are also stimulating many socioeconomic changes. It is impossible to disaggregate these broad historical processes in a way that meaningfully links China's international trade with its increasing agricultural productivity or with deforestation. It seems probable, however, that had there not been a widespread redistribution of land and other wealth in the early 1950s, and if there had not been a strong central government that depended on wide popular support in exchange for perceived social and economic development, deforestation in China would have continued after the 1950s much as it had in earlier decades. On the contrary, forested area in China has apparently increased since the 1960s, mostly as a result of extensive tree plantings. Nonetheless, this was accompanied by serious forest degradation in several regions (Rozelle, Lund, Ting and Huang, 1993).

The findings discussed earlier of the case study in Hekou county in Yunnan Province were consistent with these general observations. Large-scale deforestation took place in the late 19th and early 20th centuries when the French constructed a railroad through the area, connecting it with Vietnam's ports and cities. The resulting increased demands arising from the international trade of agricultural and forest products played an important role in the conversion of many of the region's forests to agricultural uses. These impacts were confounded, however, with rapid population growth accompanying the railroads, providing easier access to immigrants from other parts of China, and its stimulation of economic activities. Political instability and conflicts were also prevalent. Yunnan's rugged topography, however, provided a degree of protection for its remaining natural forests.

In the early 1950s there was another jump in deforestation. This was associated not with trade but with its absence. It resulted from policies during the 'great leap forward' of promoting 'backyard iron smelters' with their huge demands for charcoal and with policies aimed at making every locality self-sufficient in grain production. During this period there was almost no international trade. Subsequent expansion of forest plantations after the early 1970s and the concurrent drain of timber from many natural forests, together with continued clearance of some forest land for agriculture and other land uses, were primarily related to domestic policies and land tenure institutions.

Rapidly increasing foreign trade and investments after the 1970s made important contributions to China's spectacular economic growth in the 1980s and early 1990s. This rapid growth in turn attracted more foreign investments and trade. Most of this new international trade and investment, however, was in manufacturing and services, not in agriculture. Over four-fifths of China's imports and exports in the early 1990s consisted of manufactured goods. Linkages between international trade and deforestation in China are important, but they are, for the most part, very indirect.

## MALAYSIA

Malaysia, with three times Guatemala's area and nearly twice its population, had three times as great a GDP per capita as Guatemala in the early 1990s. It is the case study country most heavily involved in international trade in relation to the size of its economy. In the early 1990s its merchandise exports amounted to 73 per cent of its GDP while its imports were slightly greater. In the early 1990s, agricultural products comprised about one-tenth of its exports and timber exports nearly another one-tenth. In the early 1980s, they had been relatively much higher. In 1985 its exports were about one-third agricultural and forest products (principally timber, oil palm and rubber), one-third minerals (mostly tin and petroleum) and one-third manufactured products.

Malaysia has been industrializing rapidly since its independence from Britain in 1957. In 1960, agriculture accounted for three-fifths of its GDP and well over half of its labour force. By 1994 these proportions were estimated to be down to about one-fifth and just under one-quarter respectively. Exports of oil palm and rubber grew rapidly after 1960, but those of timber, minerals and especially of manufactured goods grew even faster. Malaysia is often cited as an outstanding post-Second World War example of export-led rapid economic growth.

The expansion of agriculture in colonial Malaysia in the early 20th century was stimulated primarily by tin and rubber exports. These exports required the importation of workers, largely from India, and the expansion of food crops to feed the growing workforce. Like most commodity exporters, colonial Malaysia was heavily dependent on food imports. In some periods it imported up to one-third of its rice consumption. Booming commodity exports were also accompanied by growing urban centres and important immigration from China. Forests probably covered close to nine-tenths of Peninsular Malaysia in the late 19th century. By 1946, forest cover had been reduced to about three-fourths of its land area and to about two-thirds at independence in 1957. By the late 1980s forests covered less than half of Peninsular Malaysia.

The area in rubber plantations in the peninsula expanded from nothing early in this century to nearly 1.8 million hectares in the 1970s and to over 2 million hectares in the late 1980s. Oil palm plantations were not significant until independence, but by the late 1980s they occupied over 1 million hectares. Rice and other crops also expanded, but more slowly, and their combined area was less than that in rubber and oil palm. Rubber and oil palm production was predominantly for export markets. Most deforestation in Peninsular Malaysia was directly associated with growing exports of mineral and agricultural commodities. Among these agricultural exports, timber became an increasingly important component. Much of the land cleared for rubber and oil palm expansion supported important volumes of commercially valuable tropical hardwoods. Their sale helped to defray the costs of land conversion as well as providing important additional revenues for the state, loggers, processors and many others. There was also considerable logging in forests that were not being cleared for agriculture. Log exports from Peninsular Malaysia were discouraged after the late 1970s, with the aim of encouraging its incipient wood processing industries and also of slowing the over-exploitation of its remaining forests.

Commercial logging played a secondary role to agricultural expansion in the deforestation of Peninsular Malaysia. In Sabah and Sarawak, commercial logging has been the primary process leading to deforestation. Logging interests like to blame slash-and-burn cultivators for deforestation in these states, but the evidence presented in the case study is overwhelmingly that commercial logging has been the principal process leading to recent deforestation in these two states. Its analysis of the political economy of logging showed that it was financially and politically advantageous for these states' politicians to grant large timber concessions to local entrepreneurs, frequently financed by Japanese trading companies, to extract as much timber as possible, as rapidly as possible, for export to Japan. The Japanese market for low-priced tropical hardwood was practically insatiable, much of it to be used for construction forms and cheap furniture that were soon discarded.

The states of Sabah and Sarawak, however, gained much of their revenue from the export of raw logs even when these were grossly underpriced. This was largely a result of the Malaysian institutional structure mentioned earlier, giving the federal government rights to tax-processed exports, but reserving for the states tax revenues from the export of raw timber. Also, timber concessions provided an important source of patronage for state politicians. Logging was usually wasteful and destructive. Although only a few commercially valuable trees per hectare were harvested, roads, together with heavy equipment and lack of careful logging practices, severely damaged most of the remaining stands. The rights of traditional indigenous forest users were disregarded.

There were no effective measures to ensure natural forest regeneration or replanting. Logging roads opened hitherto inaccessible forests to slash-and-burn agriculture. Japanese buyers, like commercial timber traders everywhere, were primarily interested in profits. Japan had strict regulations to ensure sustainable use of its forests at home, but its corporate trading houses were regulated only by host country law abroad. They moved from one country to another in search of cheap supplies of tropical timber.

If governments of timber importing and exporting countries could only agree on an enforceable international code of conduct setting minimal ecological and social standards for the exploitation of forest products destined for exports, this might help lessen destructive competition among low-income producer countries. The International Tropical Timber Organization (ITTO) has so far proved to be a very inadequate instrument for doing this. Meanwhile, each exporting country will have to take measures to protect its forests and its people's livelihoods.

## CAMEROON

As was seen earlier, Cameroon has an area over four times greater than Guatemala and a forest area that is nearly six times larger, but with a similar population. One-fourth of its exports in 1994 were agricultural products. In 1994 timber exports were nearly (81 per cent) as large as its agricultural exports. Forest and agricultural products together amounted to 44 per cent of Cameroon's total merchandise exports, but ostensibly to only about 6 per cent of its estimated GDP. In other words, agricultural and forestry exports were both relatively important in Cameroon's international trade, but its total international trade was a smaller component of its economy than in any of the other case study countries. (This raises questions about the data that are mentioned below.)

The export of agricultural and other primary commodities has been an important stimulus for the conversion of forests to agriculture and other uses ever since Germany made Cameroon one of its colonial possessions in the late 19th century. German investors initially cleared forests for coffee, cocoa and cotton plantations and later for rubber, oil palm, sugar and bananas. The Germans also commenced railroad construction from the coast to the eastern interior, resulting in some deforestation. After the First World War, Cameroon was divided into a British mandate in the west and a French one in the east. British and French authorities essentially continued German colonial development policies. Private investors from Britain and France acquired some of the former German plantations and also invested in new ones. After 1925 several former German investors also returned. Timber extraction for export continued but still on a rather small scale. The depression of the 1930s followed by the Second World War slowed agro-export expansion.

Following Cameroon's independence in 1960, it placed a high priority on encouraging the expansion of agro-exports as well as those of timber, oil and minerals. It also attempted to expand the country's small manufacturing sector in order to substitute nationally produced textiles and the like for imports. Most agricultural land, however, was worked by peasants who produced principally for self-provisioning and local markets.

Cameroon has long, unguarded and often unmarked frontiers with neighbouring countries, especially with Nigeria. As a result, much of its international trade is unregistered by government authorities. This helps to explain why its total merchandise trade was only 6 per cent of GDP, according to official data. In reality it was probably much higher. In addition to a great deal of unregistered transborder trade with neighbouring countries, some of which was re-exported to the North, the country's official exports of agricultural products and timber to Europe increased in the 1960s and 1970s. This was in part due to the preferential treatment given by the European Community to exports from its former colonies in Africa and the Caribbean. The largest producer of plantation crop exports, the CDC, whose activities were discussed in Chapter 3, received sizeable foreign investments. Also, private European corporations were encouraged to establish new agro-export enterprises. Cocoa, coffee, oil palm, rubber and cotton areas expanded both in modern commercial plantations and by peasant farmers who received credits from exporters. In the early 1980s, however, coffee prices fell sharply and the same happened a few years later for most other agro-exports, slowing their growth and in some cases sharply reversing it. One reason, in addition to attracting foreign capital, that the government opted for joint ventures of the CDC with foreign companies, such as Del Monte, Goodyear and Unilever and the like in agro-export production, was that the state was in a stronger position to take over land and to settle subsequent conflicts with local communities than were individual private companies.

The influence of export markets and foreign investments was particularly important in the expansion of commercial logging. This is currently the main direct source of deforestation in Cameroon. In 1991 timber exports were second to petroleum in value and the commercial timber harvest was estimated to represent 4 per cent of the country's GDP. Forest concessions are granted to national and foreign enterprises. In the early 1990s timber concessions of 4.7 million hectares to 55 foreign firms covered nearly two-thirds of the forest area licensed to be cut. Italy, France and Spain, followed by the other EU countries, and Japan were the biggest buyers of Cameroon timber.

Concessions were usually for only one cutting cycle. This left investors with no incentive for sustained yield management practices or for replanting. Harvesting practices were wasteful and damaged most remaining growing stock, especially as very heavy machinery was widely used. Also, there was no concern for the traditional rights of customary forest users. The situation was in many respects very similar to that examined earlier in Sarawak, Brazil and Guatemala.

## DIVERSE AND CHANGING LINKAGES

The case studies summarized above illustrated several linkages of international trade with agricultural expansion and tropical deforestation. They suggested that in some circumstances trade contributed to worsening livelihoods for many low-income groups while in others it helped to improve the livelihoods of people in apparently similar conditions. International trade often stimulated undesirable deforestation, but sometimes it led to more sustainable agricultural and forest land uses, depending on the context. Trade was a predominant factor driving deforestation processes in some situations but a very marginal one in others. Moreover, these relationships were constantly changing. What was a negative linkage earlier often becomes a more positive one later, or vice versa.

Of course, these diverse, contradictory and changing linkages depended on multiple factors. Country size, resource endowment and demographics influenced linkages with international markets as these links tended to be more important for small countries than for large ones, all other things being equal. Economic structure, in the sense of what was being produced, how (with what technologies), by whom (and for whom) were central determinants of trading patterns practically by definition. Economic development, where it was occurring, implied a changing economic structure that meant that some previously dominant social actors lost influence and others became more influential, but their stakes in foreign trade often differed widely.

Social institutions such as land tenure relationships were a major determinant of the impact of trade on livelihoods everywhere. Although social relationships tended to evolve slowly, they were changing in all the case study countries. Public policies also influenced patterns and impacts of international trade. These policies fluctuated wildly. So too did volatile international markets. These depended largely on policies in industrialized countries that in turn were increasingly constrained by expanding international financial and commodity markets operating within a rather chaotic world system. In each case, these and many other factors interacted in often unpredictable ways to determine actual outcomes of international trade for tropical forests and for the people associated with them.

# 5 Towards More Sustainable Use of Tropical Agricultural and Forest Resources

The area under forest cover in the tropics apparently decreased from 1910 million hectares in 1980 to 1756 million hectares in 1990. This implied an average forest loss of 15.4 million hectares annually or an annual rate of tropical deforestation of 0.8 per cent. The largest areas being deforested were in Latin America, where over half the world's remaining tropical forests exist, but the annual rate of deforestation was highest in tropical Asia. Dry tropical forests and moist deciduous forests were disappearing faster than tropical rainforests, but deforestation was advancing at between 0.6 per cent to over 1 per cent annually in all three tropical forest ecological zones in Asia, Africa and Latin America (FAO, 1993). This situation is a cause of growing national and international concern.

In this concluding chapter we summarize the principal findings of the research and its implications under four headings.

1 The causes and impacts of tropical deforestation.
2 Local level constraints and opportunities for sustainable uses of agricultural and forest resources.
3 The key role of national policies and institutions.
4 The need for international cooperation and reforms.

## Causes and Impacts of Tropical Deforestation in the Case Study Countries

All of our case studies emphasized the central roles of public policies as well as of land tenure and related institutions in stimulating (or slowing) tropical deforestation. Cross-country econometric models, on the other hand, tend to focus on such factors as population pressures, per capita income levels, investment ratios, trade intensities, relative prices and foreign indebtedness (Brown and Pearce, 1994; Palo and Mery, 1996). Some analysts highlight the importance of poverty-related agricultural expansion into tropical forests (for example, Park, 1992). Others give primacy to 'market failures' and 'policy failures' (World Bank, 1992). Focusing on public policies and social institutions in specific contexts, however, seems more likely to contribute to feasible reforms than can cross-country comparisons relating deforestation to other rather abstract and poorly measured processes.

Confusion about the causes of deforestation seems to be more rooted in epistemological and semantic problems than in an absence of empirical data. Ancient philosophical debates about the meaning of causality can never be definitively resolved. Distinguishing between proximate, precipitating and underlying causes, as is often done, can sometimes help, but it often merely adds to the confusion. There are so many interacting factors at all levels that it may be impossible to identify what role each has in a particular situation.

If one assumes that 'perfect markets' and 'good policies' would stop undesirable deforestation, then blaming it on 'market failures' and 'policy failures' is as tautological as blaming it on population increase, economic growth, poverty, agricultural expansion, trade, wasteful consumption or careless extraction of timber. Repeating these truisms in various guises is not very helpful in finding practical measures conducive to more sustainable agricultural systems and forest uses in specific contexts. The causes of unsustainable natural resource use are primarily systemic and solutions will have to include systemic reforms of institutions and policies at all levels from local to global.

Obviously, the causes of deforestation may '. . . generally originate in lands far removed from the forest' (Myers, 1994). This is why the present research focused on linkages at sub-national, national and international levels between policies and institutions that led to undesirable deforestation in specific contexts. Policies and institutions can be modified to encourage more sustainable development in tropical forest regions. We hope the information from the case studies summarized above can contribute to mobilizing social forces willing and able to bring about such reforms. This will require highly motivated special interest groups supported by much wider popular perceptions of the need for the sustainable use of tropical forests. The required policy and institutional reforms as well as the interest groups and broader popular support required to bring them about, however, will be different in each context. Moreover, unanticipated impacts of policy and institutional reforms may surprise well-intentioned initiators. This is another reason for critical analyses at all stages. But uncertainty is a poor excuse for doing nothing in the face of probable disaster.

The Brazilian government's policies attempted to expand and 'modernize' agricultural production and to extend urbanization and industrial activities into relatively sparsely populated regions of the country within an institutional structure that excluded most low-income groups from any meaningful participation. This strategy was largely responsible for accelerated deforestation in all five of the local case study areas. In each of these areas, public policies had been shaped by, and helped to reproduce, the country's land tenure system, dominated by large estates. Nearly all agricultural land was controlled by a few large owners, leaving most rural people as unstable landless rural workers or with only precarious access to small parcels of land. Government land and labour policies, credit and exchange rate subsidies, tax favours, public investments in infrastructure and a host of other measures had all contributed to outcomes that included widespread violence, decaying livelihoods and needless deforestation.

Moreover, policies ostensibly designed to counteract these undesirable social and ecological impacts of 'development' frequently ended up by reinforcing them. This was particularly true of the government's colonization policies, environmental protection policies and those supposed to protect indigenous peoples. Many well-intentioned groups helped to formulate these policies, but so too did numerous other social actors with their own agendas. The ways that they were actually applied in the case study areas were often disastrous for both the forests and the rural poor.

The accelerated clearance of Brazil's Amazonian forests since the 1950s and the mostly negative environmental and social consequences of occupying the Amazonian region were not inevitable consequences of population growth or of increased demands generated by international markets for its agricultural and forest product exports. They were a result of Brazil's social institutions and of deliberate public policies.

A similar picture to that in Brazil emerges concerning recent rapid deforestation in Guatemala. Here, however, a prolonged and bloody civil war was precipitated by the military coup of 1954. The subsequent reversal of the Arbenz administration's agrarian reform accentuated both the displacement of peasants into the forest frontier and the negative social impacts accompanying deforestation processes. The country's indigenous rural majority suffered the most, as was shown in more detail in Chapters 3 and 4.

In Guatemala the influence of international markets for its agro-exports and the trade and aid policies of the US played a much bigger direct role in stimulating recent deforestation processes than they did in Brazil. There were also greater pressures from a growing rural population. These by themselves, however, could not explain either the deforestation taking place nor the declining livelihoods of most rural residents that accompanied it. As in Brazil, profound reforms in the country's land tenure system and related institutions would be required as well as a very different and more popularly based development strategy.

China was the only case study country in which the forested area was apparently increasing at the national level (but from a very low base). Its agricultural area had hardly increased since the 1970s, while improved yields accounted for the greatest increase in agricultural production during the last two decades.

In tropical Hekou county most of the natural forests had been cleared or depleted by the mid-20th century. There was additional deforestation during the 1950s. This was primarily due to public policies associated with 'the great leap forward'. After the 1950s, better forest protection and management had been attempted by several public agencies and some had promoted plantations for timber and other forest products. Over one-fifth of Hekou county was forested in 1990, but many of these forest areas were badly depleted. Following the liberalization of economic policies and the introduction of the family responsibility system in agriculture after the 1970s, agricultural productivity had increased and tree planting had accelerated. Expansion of forest area and of cultivated land had mostly been at the expense of 'other land', which was predominantly low productivity grass lands or wasteland.

Agrarian reform in the early 1950s had provided nearly all the county's rural population with some kind of secure access to land or employment. Collectivization and state-sponsored social programmes resulted in relatively equalitarian access to available food as well as other goods and services. When the national economy grew rather rapidly in the 1980s and early 1990s, this was accompanied by somewhat improved livelihoods for the county's rural population and by relatively little new deforestation.

The institutional and policy obstacles to sustainable rural development in Hekou were associated with the poorly defined and rapidly changing rights and responsibilities in the use of farm and forest lands by different social groups. Central government, provincial, county and township agencies often had ill-defined and overlapping rights in respect to land use and forest management. In the degree that markets and price relationships became increasingly important in resource allocation, public agencies as well as peasant farmers and cooperative and other productive enterprises frequently had conflicting objectives in respect to the use of farmland and forests. It was clear, however, that forest protection could not be left primarily to decisions made in response only to market forces, without this resulting in accelerated deforestation.

Colonial policies in Malaysia were the principal stimulant for the spectacular growth of its rubber plantation production during the colonial period. They also promoted massive immigration to supply labour for tin and rubber extraction for export and related activities. Following independence, state policies encouraged the rapid expansion of oil palm plantations in hitherto forested areas. They also promoted timber exports. At the same time, state policies favoured cereal imports over rapid expansion of domestic rice production. These policies were abetted by relative prices in world markets.

Deforestation to make way for export crops virtually stopped in Peninsular Malaysia in the 1980s, but continued on a lesser scale in Sabah and Sarawak. Vast areas of Sabah's and Sarawak's forests were virtually destroyed by commercial logging for export markets. Destructive and rapid commercial timber extraction was promoted by the policies of these two island Malaysian states. These policies were encouraged by institutional arrangements that permitted each state to retain the income from timber exports but allowed the central government to tax exports of processed products. Japanese timber importers were ready collaborators with these two state governments in this deforestation.

The negative social impacts of rapid timber and agro-export growth in Malaysia were attenuated by social and economic policies that were intended to improve the livelihoods of Malaysian peasants and bring them up to near the level of the largely urban-dwelling Chinese minority. Land tenure in Malaysia was never as polarized as in the Latin American cases. New export crop plantations after independence were more participatory for their workers than colonial estates in their distribution of benefits. In Sabah and Sarawak, however, indigenous populations had little political influence and their livelihoods often declined sharply in the wake of timber exports and agro-export expansion.

In Cameroon, colonial policies had encouraged agro-export expansion that led to considerable deforestation, both directly and indirectly. They also promoted commercial timber exports. After independence, the Cameroon government continued most colonial policies favouring agro-exports and those of timber. It is difficult to estimate to what extent recent rapid deforestation in Cameroon has been a result of deliberate government policies and to what extent it has been due to the absence of effective forest protection policies by a very weak state. The end results were the same. Clearance of the country's tropical forests has proceeded at an accelerating rate. Population pressure contributed, but in lightly populated Cameroon with two-fifths of its population already urban, these were secondary factors. The local level case studies suggested that government policies of promoting agro-exports and timber exports in partnership with transnational enterprises were much more important than demographic pressures in stimulating both agricultural expansion and deforestation. A contributing factor was the setting aside of large areas in protected parks and forest reserves without adequate provision to provide alternative livelihoods for traditional users of these forest areas.

The prevalence of customary land tenure systems in spite of the government's efforts to install European-style individual property rights minimized many of the negative social impacts of agro-export expansion and forest exploitation for commercial logging. Most rural residents retained access to some land for self-provisioning, enabling them to maintain minimal livelihoods. This was less a result of deliberate policy than of the incapacity of the state to implement its declared land privatization policies.

In the light of the five case studies, it is tempting to attribute the negative social and ecological impacts associated with agricultural expansion and deforestation to 'market failures' and 'policy failures'. This is not very helpful. 'Market failure' is usually defined as the inability of market forces to produce an 'efficient' use of resources whereby marginal costs would be equal to marginal returns. This dodges the question of costs and returns for whom. Also, if prevailing (and volatile) world market prices are used as a standard, there is no reason to believe that these would necessarily be conducive to sustainable development. On the other hand, if 'shadow prices' that take into account social and environmental externalities are used to estimate market failure, the results are necessarily highly subjective, as they depend on the analysts' assessment of these extra-market values.

'Policy failure' is just as ambiguous. It assumes 'correct' policies would have a primary objective of achieving development that is socially and ecologically sustainable, but that for some reason the state failed to adopt or implement such policies. In the real world, the policies of dominant social actors, including the state itself, usually have contradictory objectives. Public policies in the case study countries in general led to consequences that many of those who promoted and executed them had intended, although the results for others were often unforeseen or unintentional. Improved livelihoods for the rural poor and sustainable use of natural resources had not been among the priorities of several of the social actors responsible for the policies. 'Policy failure' assumes a normative framework against which policies can be judged. In the real world there is no agreement about the norms of 'sustainable development' towards which policies should be directed.

The central issue is one of achieving a political consensus about what kind of development policies to promote. Scientific insights can help to illuminate constraints and opportunities but they are plagued by uncertainties. Moreover, science has little to say about who should benefit and who should pay the costs. Tropical deforestation is fundamentally a political and not a technical issue, although scientific insights can help to shape both political debates and proposed technical solutions. How deforestation narratives are told can be immensely influential in focusing discussions about the issues and in mobilizing support for possible initiatives to confront them.

## LOCAL-LEVEL CONSTRAINTS AND OPPORTUNITIES

Most environmental problems are local in the sense that their sources tend to be site-specific and that their negative impacts are largely borne by local people, usually the poorest, and by local ecosystems. Their solutions require local actions with the participation of local people. The popular environmentalist slogan of 'think globally and act locally', however, can be very misleading if it leads to the neglect of the supportive regional, national and international policies and institutions required for successful local initiatives towards more sustainable agriculture and forestry. Local efforts to control undesirable deforestation face numerous insurmountable constraints unless complemented by policy and institutional reforms at higher levels, as the case studies reviewed above make abundantly clear.

There are always actions that could be best taken at local levels to promote more sustainable agriculture and forestry. In some circumstances there are wide margins for local initiatives while in others these margins are extremely narrow. Local initiatives can be conducive to more sustainable use of agricultural and forest resources, but they can also be detrimental. This depends largely on local power structures as well as the objectives and perceptions of local élites. Decentralization by itself is no panacea for approaching sustainable development and it can have negative consequences in many settings. Truly democratic and participatory local governance, however, could help in most situations.

Opportunities for local initiatives towards sustainable development often lie with intensifying agricultural production using available labour and a combination of traditional low-external-input farming practices enhanced by selective technical improvements and inputs associated with modern science. More sustainable farming systems may be quite impossible, however, in situations where local people have no secure control over their resources and where their products are virtually confiscated by others such as landlords, middlemen, political and military authorities or mere brigands. In addition to such all too common political constraints, local people have to confront constraints imposed by an often hostile physical environment.

The success or failure of local initiatives have to be judged both by their longer term impact on the environment and on the livelihoods of diverse groups and strata of low-income people in each locality. What may look like a success after three or four years may appear to be a failure after one or two decades, and one that appears to fail in the short run may look successful a few decades later. Much depends on external factors in either case. Moreover, an initiative that benefits one low-income group may prejudice others. Forests may be saved in one place through strict protection only at the cost of clearance of neighbouring forests by traditional users of the protected area in search of new livelihoods.

The case studies in Brazil and Guatemala documented many instances of poor peasant communities attempting to defend their insecure control of agricultural and forest lands that they had been using in ecologically sustainable farming systems. They usually lost their battles. Often they were violently evicted or worse. Occasionally they gained at least a temporary victory. When they did, support from outsiders such as national or international NGOs who helped them find allies from outside of the locality was usually crucial. NGOs and public agency employees sometimes helped to introduce new low-cost agricultural practices that enabled peasants to increase their yields. The scope for improving livelihoods through improved farming and forestry practices, access to some credit, better marketing and new economic activities, however, was nearly always constrained by precarious or insufficient access to land.

Perhaps one of the most important functions outsiders interested in promoting more sustainable development at local levels can have is to act as a catalyst in stimulating local groups of poor peasants and workers to try to overcome some of the socioeconomic and political constraints that prevent them from improving their livelihoods. In the Ribeira basin case study area in Brazil there had been some signs that increasingly vocal peasant organizations might have some influence in modifying public policies detrimental for both them and their natural environment. This was in relatively rich and democratic São Paulo state, where there were potentially powerful urban-based allies. Similar initiatives in several Amazon states and in rural Guatemala had been violently repressed. When repression occurred, it was usually the local poor and not their NGO or other allies from elsewhere who paid the costs. Outsiders have to use very keen and well-informed judgement in promoting organization of the poor to improve their terms of access to natural resources. They may unintentionally contribute to reprisals such as harsh repression. Also, the presence of outside organizations in any locality usually depends on the acquiescence of national governments.

Democratic local institutions that are somehow accountable to the poor as well as to élites are prerequisites for approaching sustainable agriculture and forestry. In their absence, there is no possibility of achieving real and lasting participation of local people. Elite groups primarily accountable only to their peers and to outside interests or authorities cannot be expected to make the sustainable use of natural resources for the long-term benefit of the local population their top priority. Local authorities who are primarily accountable to agencies or corporations based elsewhere, that in turn do not have to maintain the goodwill and support of the local population, tend to act in a very arbitrary fashion when dealing with peasants and forest dwellers. Administrators of protected forest areas in the Ribeira basin, for example, rigidly applied regulations prohibiting peasants from cutting any trees or cultivating land in the forest areas they had been using for generations in a sustainable manner. This kind of forest protection becomes a kind of eco-fascism. It is not socially sustainable.

Most of the political and economic constraints on the sustainable use of natural resources by people in a rural locality originate elsewhere. They can seldom be removed through local initiatives alone. Secure and equitable access to land, for example, is no longer purely a local matter practically anywhere. Even where customary communal tenure still prevails, such as in much of rural

Cameroon, peasants may be deprived of their rights arbitrarily and capriciously by the state or corporate enterprises.

Without the mobilization and organization of peasants and other low-income rural residents with the aim of gaining greater control over resources and institutions, however, there seems little likelihood that wider sub-national, national and international political systems will seriously take their interests into account when formulating and executing policies affecting them. Local-level initiatives aimed at approaching more sustainable development through community-based resource management necessarily imply a struggle for greater control over resources and institutions in specific social contexts by those hitherto excluded from such control. Such struggles for self-empowerment are inevitably highly conflictive.

## The Crucial Role of National Policies and Institutions

Many observers have been celebrating or lamenting the decline of the national state, which they see as atrophying with the advance of globalization. Nonetheless, the state remains the principal agent potentially capable of bringing about more sustainable patterns of development. At least in theory, it retains a monopoly of the legitimate use of force in its territory. It is responsible for formalizing and enforcing the legal rules under which the national society operates, including those governing the use of its natural resources and the distribution of their benefits. Only the state is recognized under international conventions to have the legitimate right to redistribute wealth and income in 'the public interest'. The state is the final arbiter of legal disputes within its jurisdiction. It is also expected to have a pre-eminent role in determining national macro-economic and social policies. International organizations of all kinds from intergovernmental bodies to transnational corporations ultimately depend on nation states for their own legitimacy as well as to enforce rules and contracts. The World Bank seems to have rediscovered the central importance of the nation state in guiding economic and social development after two decades of preaching that these issues could largely be better left for private agents and market forces to resolve (World Bank, 1997).

As noted earlier in this chapter, the case studies all highlighted the central importance of national policies and institutions in shaping agricultural expansion and deforestation processes as well as in determining their impacts on livelihoods and on the environment. While the research focused attention on agrarian and forest-related policies and institutions, it emphasized that these were only components of dominant national development strategies. Macro-economic policies affecting growth and employment and social institutions determining the distribution of costs and benefits were always of central importance. Piecemeal policies aimed at promoting more sustainable agriculture and forestry were often ineffective or counterproductive because they were not supported by complementary policies in the broader society.

National strategies that are sustainable, however, have to be popularly based in the sense that those formulating and executing them perceive that they are somehow accountable to low-income constituents for the way these policies affect livelihoods both in the immediate future and in the longer term. The Brazilian case in particular brought out how policies sincerely intended by some of their sponsors to help the rural poor as well as to protect indigenous groups and forest ecosystems, were in practice subverted to serve the short-term interests of powerful state support groups with very different objectives.

How to bring about and maintain popularly based national strategies directed towards development that is socially and ecologically sustainable is a central issue everywhere. For this to happen, those wielding state power have to perceive the diverse social groups that constitute their countries' low-income majorities as potentially crucial allies or troublesome opponents likely to become allies of competing élites. This implies that organized and vocal pressures from the poor are indispensable for the emergence of a popularly based sustainable development strategy. But this is not sufficient in itself. Poor people have to find more powerful allies who have conflicting agendas. Peasants are usually well aware of the importance of maintaining the productivity of their agricultural lands and of their forest resources to meet their immediate needs as well as those of future generations. They cannot be expected, however, to mobilize and organize around such abstract environmental issues as global climate change and loss of biodiversity unless these are perceived in terms of their own and their children's livelihoods. Also, the urban poor face many different environmental problems than do their rural counterparts. In addition, the political system often provides few or no channels whereby low-income groups can articulate their needs in ways that are heeded by the state.

Forging durable alliances leading to national development strategies that are socially and ecologically sustainable is always a formidable, and perhaps an impossible, task. Statesmen, political and civic leaders, educators, scientists and just about everyone else has to become constructively involved. Those NGOs that are primarily concerned with human rights, socioeconomic development and ecological issues can often help by playing a catalytic role. But their efforts can easily become dissipated or even harmful if by focusing exclusively on a single issue they neglect the complexity of the central problem and the possible negative effects of the solutions they propose for many low-income social groups. The establishment of strictly protected forest areas without due consideration of the harm these may cause to the livelihoods of customary users of these forests is a typical example (Ghimire and Pimbert, 1997).

In-depth 'political economy'-style analyses are required to identify the crucial social actors and their interests, as well as mobilizing the broader popular support and alliances needed to bring about more sustainable development strategies. Answers at best would have to be speculative and different for each situation (Barraclough and Ghimire, 1995). The case studies, however, were able to identify several specific policies and institutions in each country that were apparently directly encouraging undesirable deforestation while negatively affecting the livelihoods of many low-income residents of tropical forest regions. They also suggested some possible reforms that might help bring about more sustainable development in these areas.

In Brazil and Guatemala land tenure institutions were a fundamental issue for the rural poor. The control of most agricultural land was monopolized by a small group of large landholders, while access to adequate land was denied to most rural people.

The rights and obligations associated with land ownership and other forms of land and forest tenure were ill-defined and precarious for the poor as well as being highly inequitable. Small producers could easily be evicted and their lands appropriated without compensation for the profit of large holders, speculators, urban based investors and often the state itself. Displaced peasants and landless workers usually had no alternatives other than migration to the cities or to the forest frontier in order to survive. Ample good agricultural land was theoretically available in both countries, especially in Brazil, to have provided their entire rural populations with access to enough land to meet their needs for self-provisioning as well as production of a surplus for domestic and export markets. In these two countries, redistributive land reforms clearly should have a high priority in any popularly based strategy aimed at sustainable development.

The land rights of Guatemala's indigenous majority and of Brazil's small indigenous minority were particularly precarious. These indigenous peoples had developed ecologically sustainable farming systems that were well adapted to their tropical forest environment. Their customary farming systems became unsustainable when indigenous lands were appropriated for agro-industrial monocropping, cattle ranching, commercial logging, nature reserves and other uses. Agrarian reform would have to provide special treatment in respect to indigenous land rights in order not to provoke new injustices in the name of development.

In any case, land reform would have to be adapted to each situation and complemented by many other policies and institutional changes in the broader society in order to be effective in facilitating more sustainable natural resource management while improving peasants' livelihoods. For example, to be effective it would have to be associated with appropriate fiscal, price, trade, investment and employment policies. Research, technical assistance and credit priorities would have to be revised and there would have to be profound reforms of political and judicial institutions. These are issues that have to be dealt with politically within each country, taking into account its particular situation. There is little to be gained from outsiders offering their advice in general terms. A more supportive international policy and institutional environment, however, could help in stimulating national reforms.

Inequitable and precarious land tenure rights in lightly settled rural Cameroon seemed likely to become a serious obstacle to the emergence of sustainable farming and forestry systems in response to the government's drive to increase agro- and timber exports while 'modernizing' production technologies. The negative impacts of agro-export expansion for rural livelihoods had been attenuated by the persistence of customary land systems. This occurred in spite of state policies aimed at replacing traditional common property régimes with private control of most agricultural land and state control of forest resources. It is not easy for a weak state to impose unpopular policies on reluctant rural populations. There is a real danger, however, that Cameroon's land tenure system will soon become as inequitable and precarious for most rural people as that of Brazil or Guatemala. Considerable areas have already been alienated from customary users for agro-export crop production and for strictly protected parks and nature reserves. Many forest areas are being carelessly exploited under state-granted concessions for commercial timber exports. As happened earlier in South Africa, Zimbabwe, highland Kenya and many other regions of Africa, bimodal land tenure systems (characterized by land being mainly held by a few large estates, with most farmers having only very small holdings) are evolving resembling those that have dominated Latin America.

Cameroon's trade and investment policies have promoted agro- and timber exports with little concern for socially and environmentally sustainable national development. Agro-export crops have expanded but the growing food needs of its rapidly urbanizing population are, to a large extent, met by subsidized food imports from Europe. Its peasants have few incentives or access to the credit and inputs required to produce food surpluses for the cities. Its forest resources are being depleted for timber exports. The state does not have a coherent strategy for encouraging sustainable development.

In Malaysia, land tenure rights tended to become somewhat more equitable and secure for many of the rural poor following independence. The state perceived an urgent need to adopt policies designed to improve the livelihoods of Malaysian peasants and rural workers in order to maintain their crucial political support, first in the face of communist insurgency and later to defuse the appeal of rival political élites competing for state power. Indigenous groups in Sabah and Sarawak, however, lacked sufficient political weight for the protection of their traditional land rights to be a serious priority for the state government or for the federal government.

A leading institutional issue concerning land rights directly affecting current tropical deforestation in Malaysia is the division of revenues from forest exploitation between the state and federal governments. As long as the state governments of Sabah and Sarawak must depend on income from exports of raw timber for a substantial part of their revenues and political patronage, and as long as Japanese and other transnational companies will purchase their still unexploited timber resources for short-term gains, it seems unlikely that these state governments will adopt policies conducive to sustainable forest management.

China's drastic land reform of the early 1950s, together with several complementary social and economic policies, contributed to providing a foundation for a later, more popularly based and sustainable development strategy. Some of the revolutionary government's policies such as those of promoting backyard iron smelters and local-level self-sufficiency in grains during the 'great leap forward' accelerated deforestation. On the other hand, more equitable and secure access to land by the peasant majority and public policies intended to improve access by the rural poor to health and educational services as well as to provide them with a more equitable access to available food and consumer goods slowed the invasion of forest areas by destitute peasants in desperate search of minimal livelihoods.

The policy reforms of the late 1970s and early 1980s introducing the 'family responsibility' system in agriculture and a greater role for market forces in resource allocation in general were conducive to more rapid industrialization and economic growth. Employment opportunities outside of agriculture increased rapidly in both rural and urban areas. In spite of the growing population of the world's most populated, and one of its poorest, countries, the area under forest cover apparently increased slightly. Agricultural production from a nearly stable agricultural area grew faster than its population, although not enough to meet all its rapidly growing demands for food and other farm products.

Whether China's current development pattern can, on balance, become sustainable is a matter of debate. In any case, the frequent changes in policies during the last five decades can be better analysed as successive steps in a complex historical process that has enjoyed considerable widespread popular support than as the mere substitution of policies that failed by others that were more successful. Usually the success of new policies depended at least in part on the foundations laid by the earlier ones that they replaced.

The case study of Hekou county suggested the desirability for clearer definitions of the rights and responsibilities of different public, cooperative and private entities at national, provincial and sub-provincial levels in the use of agricultural and forest resources. Conflicting objectives and jurisdictions often led to unsustainable resource management. Many of these agents were more concerned with short-term financial profits than with long-term sustainability. This was largely because the structure of rewards and penalties provided by the institutional framework failed to account adequately for many externalities. There has been considerable underinvestment in sustainable high productivity forest management as a result. Also, some agricultural expansion into lands more suited for forests could have been avoided by more emphasis on increasing yields of agricultural areas using improved but ecologically sustainable farming practices. On the whole, however, short-term prospects for reversing undesirable tropical deforestation in low-income Hekou county appeared brighter than in low-income Cameroon, or in much-higher-income Brazil and Guatemala, or in Sabah and Sarawak in far-higher-income Malaysia.

Deforestation in tropical China had already proceeded much further than in these other countries before the state had reacted with a popularly based strategy aimed in part at achieving more sustainable use of its severely limited agricultural and rapidly dwindling forest resources. There usually has to be widespread recognition that serious and urgent social and environmental problems exist and that solutions are possible through collective actions before much can be done about them through public policies.

This discussion of national policies and institutions has emphasized land tenure-related issues because they are so fundamental, especially in predominantly agrarian countries. They reflect both the distribution of power among diverse social groups and the difficulties of changing these relations to empower those whose survival depends on the terms under which they can obtain access to land and other natural resources. The policy and institutional reforms required for sustainable development, however, extend far beyond those dealing with access to land and other natural resources. The whole complex of institutions and policies comprising national societies has to adapt to meet the need for sustainable development, as the case studies make abundantly clear.

All the case studies suggested that undesirable tropical deforestation due to human activities is primarily a result of deliberate political choices. It is not caused by uncontrollable socioeconomic and demographic forces. Policies and institutions can be reformed to encourage development that is socially and ecologically sustainable. The nation state remains the principal agency through which such reforms can potentially be brought about and made effective. The challenge is to identify and mobilize the social forces willing and able to move the state in this direction. To be effective, they will have to include the active participation of popularly based organizations such as labour unions, peasant leagues, consumers' associations, indigenous communities and the like. These groups potentially represent the interests of low-income majorities who stand to gain the most from sustainable development.

## INTERNATIONAL REFORMS

Tropical deforestation cannot be expected to stop when a country reaches a certain level of affluence.[1] The sustainable use of agricultural and forest resources requires policy and institutional reforms at all levels and especially by the nation state. There are several international reforms that could support and, in some

1 This is borne out by the case studies. Deforestation trends in high-middle-income Brazil and Malaysia was more serious than in much-lower-income China and certainly no less a problem than in lower-middle-income Guatemala or Cameroon. One econometric analysis of data from 66 countries concluded that 'statistically speaking, per capita income has virtually no explanatory power (for deforestation) ...' (Shafik, 1994).

cases, stimulate needed reforms at national and sub-national levels (Barraclough, Ghimire and Meliczek, 1997). The research summarized in this book did not focus on international policies and institutions. Linkages observed at local and national levels, however, enable us to speculate about a few international reforms that could help.

The rich industrial states, together with the international financial institutions that they control, have not usually supported the adoption of popularly based strategies of sustainable development in poor countries. Their insistence on structural adjustment with rigid monetary, fiscal, trade and privatization policies that conform with neo-liberal criteria has often discouraged developing countries from adopting socially and ecologically friendly strategies. The burden of servicing heavy foreign debts to the North is a major obstacle for financing social and environmental programmes in the South. Trade and immigration restrictions in the North are counterposed to its insistence on trade liberalization in the South and its granting unrestricted access to Northern investors. The North has also consistently refused serious negotiations aimed at stabilizing volatile prices of Southern commodity exports and employing Northern gains from longer term deteriorating terms of trade for commodity exporters to help finance efforts towards more sustainable development in poor countries.

Without substantial progress towards a more democratic world system, sustainable development will remain elusive everywhere (South Centre, 1996). Finding and mobilizing the social forces required for a strong and democratic United Nations system capable of steering the world economy towards more sustainable development should be high priority for anyone concerned about tropical deforestation. National governments that are themselves democratic and popularly based will have to take the lead. Peoples' organizations and NGOs can also play an important role.

The negotiation of international codes or agreements setting minimum social and environmental standards for transnational corporations and investors should have a high priority on the international agenda. International norms, however, are likely to be unenforceable and even counterproductive if they are imposed by rich states on poor ones without being genuinely accepted by the latter and by the poor majorities of their inhabitants. Such standards would have to take into account social and environmental externalities according to the 'polluter pays' principle. Many poor country governments have argued that this would undercut their competitive advantages derived from very low wages, unsustainable exploitation of natural resources and lax environmental regulations. This does not have to be the case. International codes will have to be democratically formulated with the full participation of representatives of poor majorities from both North and South. Minimum standards could benefit poor countries as a group by preventing cut-throat competition among themselves to offer transnational investors the best possible terms even if these mean overexploitation of both their people and their natural resources. Collectively they could insist upon a better deal that in the longer term would be to all poor countries' benefit. In any event, if protectionist pressures grow in the rich countries, they would not need to invoke social and environmental standards in order to raise their barriers to exports from the South, as they could easily find other excuses.

In the face of accelerating integration of the world economy under the aegis of transnational corporate entities, greater international regulation of transnational trade, investment and financial markets is inevitable. The danger is that the rules will be made by the transnationals themselves, in cooperation with rich country governments, for their own short-term gains. Already some transnational enterprises are calling for international environmental standards that would depend on production processes over which they have virtual monopoly control of the necessary technologies. This would freeze out competitors, especially those from low-income countries. The proposed multinational agreements on investment (MAIs) being pushed by the OECD and the WTO have been drafted to serve the interests of transnational investors by increasing their freedom but omitting to specify their responsibilities. Moreover, treating foreign investors as if they were nationals of countries in which they invest would severely constrain developing countries trying to pursue strategies designed to promote self-reliant and sustainable national economies. Proposals to regulate international financial markets made by some of the world's richest financiers would make the regulators accountable to central bankers, not to democratically constituted international institutions.

International codes prepared by the rich to protect their own interests are not likely to place a high priority on achieving greater social justice either now or for future generations. The need for international regulation is becoming imperative, but the issues of who makes the rules and whose interests are served are crucial for sustainable development. The interests of poor people in poor countries will inevitably be neglected unless they have an influential voice in creating the regulatory framework and unless the regulators are accountable to low-income groups as well as to the wealthy corporate bodies that could also benefit from prudent rules and greater transparency.

Transnational investors should be required to finance participatory social and environmental impact assessments for projects they propose to finance in developing countries. Projects that fail to meet minimum social and environmental standards would be ineligible for publicly-supported international funding. Enforcement of international standards would be difficult or impossible in most countries, but their mere existence would help peoples' organizations, NGOs and others striving for a more sustainable development to mobilize opposition to harmful or dubious projects.

The 'polluter pays' principle should be an integral part of such international norms. Social and environmental criteria should be used to estimate the implicit costs and benefits for diverse social groups affected by investments and other activities with significant impacts on the use of forests and other natural resources. Improvements are required in preparing national accounts and in project evaluations in order to emphasize these externalities. Such estimates would have to be heavily qualified, however. It is impossible to estimate objectively in monetary terms the social and environmental costs and benefits associated with alternative development strategies and policies. They cannot be compared meaningfully using any single scale of value such as money or energy. Their impacts for different social groups with diverse values are inherently incommensurable (Martinez-Alier, 1989). Moreover, all such estimates are subject to huge uncertainties. Nonetheless, monetary estimates can be useful in specific contexts for designing taxincentives or penalties and related measures to encourage more sustainable practices. They could also help in convincing taxpayers in rich countries to support higher levels of funding in support of programmes aimed at promoting sustainable uses of the South's tropical forests.

International standards, codes of conduct, pollution and energy taxes, subsidies, accounting and similar reforms cannot by themselves bring about sustainable development in tropical forest regions. This is primarily a problem that has to be dealt with nationally and sub-nationally in each country. An international system that encourages popularly based national strategies aimed at promoting socially and ecologically sustainable development in general and in tropical forest regions in particular could facilitate the adoption of desirable policy and institutional reforms. Some of these proposed international reforms would require greatly increased financial resources that would have to be primarily raised in rich countries. The costs for the rich of supporting sound initiatives towards more sustainable development now could be incalculably smaller than the costs that would probably fall on them and their descendants if current trends were to continue.

The evidence reviewed above showed that agricultural expansion and international trade were linked to tropical deforestation processes in all the case study countries.[2] These linkages, however, were different and frequently contradictory for each place and time. Moreover, they were always mere components of interacting policies and institutions at all levels. International policy and institutional reforms could help in enabling and stimulating needed reforms at national and sub-national levels, but they cannot substitute for them. The road towards more sustainable development is much too complex and uncertain to be found by simplistic approaches internationally, nationally or locally.

The rich industrial countries will have to take the lead in confronting social and environmental issues on a global scale. Any international standards will have to include the rich countries as

2 Better and more detailed disaggregated data on land use trends in developing countries could help in understanding these linkages. As was seen in Chapter 2, the FAO's and other international organizations' data are woefully inadequate. Improvement of information about land use trends and the processes that are driving them is expensive and time-consuming. Moreover, it is not likely to help much in bringing about needed policy and institutional reforms unless the relevant social actors at local and sub-national levels are all deeply involved in generating them.

well as poor ones. The North should not expect developing countries to give up sovereignty over their natural resources such as tropical forests unless the rich countries are willing to do the same. Developing countries will have to take primary responsibility for dealing with their own social and environmental problems, but they need a more supportive international context.

National and international initiatives to protect rural livelihoods and the environment are doomed to be ineffective if they do not confront the fundamental social issues generating unsustainable, inequitable growth. A truly participatory international effort at all levels is imperative. The key issue remains that of what social actors might bring about the required institutional and policy reforms for the improvement of rural livelihoods while at the same time protecting tropical forests for present and future generations.

# Bibliography


Angelo-Furland, Sueli and Francisco A De Arruda Sampaio (1995) *Government Policies, Agriculture, and Deforestation in Brazil: An Introductory Approach through Five Case Studies,* Instituto de Pesquisas Ambientais: São Paulo

Arruda, Rinaldo SV (1995) *Políticas públicas, agricultura e desmatamento no Brasil. Estudo de caso 3: Os Rikbaktsa e a conservação dos recursos naturais,* Instituto de Pesquisas Ambientais: São Paulo

Barraclough, Solon and Krishna Ghimire (1995) *Forests and Livelihoods: The Social Dynamics of Deforestation in Developing Countries,* Macmillan: London

Barraclough, Solon, Krishna Ghimire and Hans Meliczek (1997) *Rural Development and the Environment: Towards Ecologically and Socially Sustainable Development in Rural Areas,* UNRISD: Geneva

Brown, Katrina and David W Pearce (eds) (1994) *The Causes of Tropical Deforestation: The Economic and Statistical Analysis of Factors Giving Rise to the Loss of the Tropical Forests,* CSERGE, University of East Anglia; University College London, UCL Press: London

Bulmer-Thomas, Victor (1994) *The Economic History of Latin America Since Independence,* Cambridge Latin American Studies, Cambridge University Press: Cambridge

Castro Oliveira, Bernardete AC (1995) *Políticas públicas, agricultura e desmatamento no Brasil, Estudo de caso 4: Posseiros e Agropecuárias na Amazônia Legal: o caso de Mirassolzinho – Jahuru, MT,* Instituto de Pesquisas Ambientais: São Paulo

FAO (Food and Agriculture Organization of the United Nations) (1988) *An Interim Report on the State of Forest Resources in the Developing Countries,* FAO: Rome

FAO (1991) *Second Interim Report on the State of Tropical Forests,* Forest Resource Assessment 1990 Project, mimeo, paper presented at the 10th World Forestry Congress, FAO: Paris

FAO (1993) *Forest Resources Assessment 1990: Tropical Countries,* FAO Forestry Paper 112, FAO: Rome

FAO (various years) *Production Yearbooks,* for the years *1958, 1959, 1960, 1961, 1969, 1970, 1973–1993* and *1995,* FAO: Rome

García, Rolando (1984) *Food Systems and Society: A Conceptual and Methodological Challenge,* UNRISD Food Systems Monograph, UNRISD: Geneva

Ghimire, Krishna (1994) *Conservation and Social Development: A Study Based on an Assessment of Wolong and other Panda Reserves in China,* Discussion Paper No. 56, UNRISD: Geneva, December

Ghimire, Krishna B and Michel P Pimbert (eds) (1997) *Social Change & Conservation: Environmental Politics and Impacts of National Parks and Protected Areas,* Earthscan Publications Ltd: London

He, Bochuan (1991) *China on the Edge: The Crisis of Ecology and Development,* China Books & Periodicals, Inc: San Francisco

Hecht, S and A Cockburn (1990) *The Fate of the Forest,* Penguin: London

Heilig, Gerhard K (1997) 'Anthropogenic factors in land-use change in China', *Population and Development Review,* Vol 23, No 1, March, pp139–168

Holland, M, R Allen, K Campbell, R Grimble and T Stickings (1992) *Natural and Human Resource Studies and Land Use Options: Department of Nyong and So'o, Cameroon,* NRI: Chatham, UK

Li Jinchang, Kong Fanwen, He Naihui and Lester Ross (1987) 'Price and policy: The keys to revamping China's forest reserves', in R Repetto and M Gillis (eds), *Public Policies and the Misuse of Forest Resources,* Cambridge University Press: Cambridge

Martinez-Alier, J (1989) *Ecological Economics,* Blackwell: Oxford

Meadows, DH, DL Meadows, J Randers and WW Behrens III (1972) *The Limits to Growth, A report for the Club of Rome's Project on the Predicament of Mankind,* A Potomac Associates Book, Earth Island Ltd: London

Menzies, N and N Peluso (1991) 'Rights of access to upland forest resources in southwest China', *Journal of World Forest Resource Management,* Vol 6, pp1–20

Ministry of Planning and Regional Development, WWF, Commission of the European Commission and Overseas Development Natural Resources Institute (undated) *Republic of Cameroon, The Korup Project,* London

MOA (Ministry of Agriculture) (1991) *Sustainable Agriculture and Rural Development in China: Policy and Plan,* Document of Ministry of Agriculture: Beijing

Mope Simo, JA (1995) *Agricultural Expansion and Tropical Deforestation in Cameroon* (Second Draft Report), June, commissioned by UNRISD, Geneva and WWF International: Gland

Myers, N (1989) *Deforestation Rates in Tropical Forests and their Climatic Implications,* Friends of the Earth: London

Myers, N (1994) 'Tropical deforestation: rates and patterns', in Katrina Brown and David W Pearce (eds), op cit

de Oliveira, Ariovaldo Umbelino (1995) *Políticas públicas, agricultura e desmatamento no Brasil. Estudo de caso 1: Assentamentos humanos na amazônia matogrossense: O caso de São Felix do Araguaia,* Instituto de Pesquisas Ambientais: São Paulo

Palo, Matti and Gerardo Mery (1996) *Sustainable Forestry Challenges for Developing Countries,* Kluwer Academic Publishers: London

Park, Chris C (1992) *Tropical Rainforests,* Routledge: London

QGLYTJHB (1989) *Quanguo Linye Tongji Huibian, 1949–1987* (National Compendium of Forestry Statistics), Ministry of Forestry Press: Beijing

Roberts, Neil (1996) 'The human transformation of the Earth's surface', *International Social Science Journal. Geography: State of the Art I – The Environmental Dimension,* UNESCO, December, pp493–510

Ross, L (1988) *Environmental Policy in China,* Indiana University Press: Bloomington, IN

Rozelle, Scott, Heidi Albers and Li Guo (1995) *Forest Resources Under Economic Reform: The Responses in China's State and Collective Management Regimes,* Stanford University: Stanford

Rozelle, Scott, Susan Lund, Zuo Ting and Jikun Huang (1993) *Rural Policy and Forest Degradation in China*, Stanford University: Stanford, 23 August

Sader, Regina (1995) *Agricultural Expansion and Tropical Deforestation in Brazil. Case Study 2: The Case of 'Bico do Papagaio',* Instituto de Pesquisas Ambientais: São Paulo

Shafik, Nemat (1994) 'Macroeconomic causes of deforestation: barking up the wrong tree?', in Katrina Brown and David W Pearce (eds), op cit

Shunwu, Zhou (1992) *China Provincial Geography,* Foreign Languages Press: Beijing

Simo, Mope (1995) *Agricultural Expansion and Tropical Deforestation in Cameroon* (second draft report), UNRISD: Geneva

South Centre (1996) *Universal Food Security: Issues for the South,* South Centre: Geneva

Smil, V (1984) *The Bad Earth: Environmental Degradation in China,* ME Sharpe, Inc: New York

Sundaram, Jomo and Chang Yii Tan (1994) *Agricultural Expansion and Deforestation in Malaysia* (draft), UNRISD: Geneva

UNDP (United Nations Development Programme) (1994) *Human Development Report 1994,* Oxford University Press: Oxford

UNDP (1996) *Human Development Report 1996,* Oxford University Press: Oxford

Utting, Peter (1993) *Trees, People and Power: Social Dimensions of Deforestation and Forest Protection in Central America,* Earthscan Publications Ltd: London

Valenzuela, Ileana (1996) *Agricultura y Bosque en Guatemala: Estudio de caso en Petén y Sierra de las Minas,* UNRISD, Geneva; WWF, Gland; and Universidad Rafael Landívar, Guatemala City

World Bank (1991) *Country Study on the Forestry Subsector of Malaysia,* World Bank

World Bank (1992) *World Development Report 1992,* Oxford University Press: New York

World Bank (1997) *World Development Report 1997,* Oxford University Press: New York

WRI (World Resources Institute) (1988) *World Resources,* Oxford University Press and WRI: New York

WRI (1990) *World Resources 1990–91,* Oxford University Press: New York

WWF (World Wide Fund For Nature) (1989) *Tropical Forestry Conservation,* a WWF International Position Paper, No 3: Gland, August

ZGTJNJ (1992) *Zhongguo Tongji Nianjian* (China Statistical Yearbook), China Statistical Press: Beijing

Zuo, Ting (1993) '*Sustainability of Rice Cropping Systems in Mountainous Yunnan, China*', paper presented at the Second Workshop of the China Rice Economy Project, China National Rice Research Institute: Hangzhou, China

# Index